Transdisciplinary Environmental Humanities

Volume 1

Series Editors
Rod Giblett
Deakin University
Coburg, VIC, Australia

Rachel Fetherston
Deakin University
Bowenvale, VIC, Australia

In a time of climate catastrophe, mass species extinction, the end of the age of fossil fuels, and the recent resurgence of reactionary forces against progressive politics, this series presents transformative studies of these and other environmental topics from across the arts, humanities and sciences. Central to the series is the nexus between the human-nature and the human-technology relation, providing critical theories of technics, discourse and fiction with a fulcrum point for developing new ways of thinking and imagining in an increasingly screened world ruled by the algorithms of computational capitalism.

The series proposes ways for living with and caring for nature locally, globally, bio-regionally, socially, and psycho-symbiotically in an increasingly networked world. It does so according to principles of justice and morality tested by critique and eco-thinking that approaches the earth's various human and more-than-human spheres as deeply interwoven and relational rather than discrete.

This series is transnational as it crosses global and cultural divides. It is trans-species as it embraces human and more-than-human beings. It engages in cross-cultural and transhistorical dialogue between cultures. It is trans-geographical as it affirms the unique worth of places, engages in decolonial critique of landscapes of culture and nature, and demonstrates their intertwining in livelihoods across the globe. It is trans-generic as it traverses the divide between fiction and non-fiction through transdisciplinary studies of a wide variety of texts, events and mediated milieus, including but not limited to literature, film, television, games, theatre, social media, and the screened realities of networked life. It is transdisciplinary within the arts and humanities and across the great divide between 'the two cultures' of the sciences and the humanities on topics of common concern for both the human and more-than-human. Proposals for monographs, edited volumes, and shorter works (Palgrave Pivots) are welcome.

Rod Giblett

The Politics of Nature and Environmental Writing

Political Ecocriticism

palgrave
macmillan

Rod Giblett
Coburg, VIC, Australia

ISSN 3091-3993 ISSN 3091-4000 (electronic)
Transdisciplinary Environmental Humanities
ISBN 978-3-032-04426-6 ISBN 978-3-032-04427-3 (eBook)
https://doi.org/10.1007/978-3-032-04427-3

This Palgrave Macmillan imprint is published by the registered company Springer Nature Switzerland AG.
The registered company address is: Gewerbestrasse 11, 6330 Cham, Switzerland

*To the conservation of Swallowtail butterflies (*Papilio machaon Britannicus*),*
*their larval foodplant Milk-parsley (*Thyselium palustre, *Marsh thyme),*
and their only home in the wetlands of Norfolk, England

Acknowledgements

I am grateful to Iain H. Leach for providing gratis a high-resolution version of his photograph of a Swallowtail Butterfly (*Papilio machaon Britannicus*) and Milk-parsley (*Thyselium palustre*, Marsh thyme), its larval foodplant in a Norfolk wetland. I am also grateful to Iain for his permission to reproduce his photo on the cover of *The Politics of Nature and Environmental Writing*. Nature and environmental writing often focus on a charismatic and endangered species and its equally endangered, but often non-charismatic, habitat. I follow suit on both scores here in this instance. The scientific name of the larval foodplant *Thyselium palustre*, literally 'Marsh thyme,' makes the connection with the paludal that I have discussed recently in *Wetland Cultures: Ancient, Traditional, Contemporary* (Palgrave Macmillan, 2024). I am also grateful to Patrick Barkham for his article about the Swallowtail Butterfly in *The Guardian* in April 2024 that drew my attention to it, its larval foodplant, its home habitat of wetlands, and Iain's photograph of butterfly and foodplant. I recommend contemplating the fragile beauty and precarious plight of the Swallowtail Butterfly and their home habitat to the accompaniment of Robert Schumann's *Papillons* opus 2 or Dora Pejačević's *Papillon* opus 6.

I am grateful to various friends, colleagues, and reviewers for their helpful suggestions of references for me to follow up and for their equally helpful comments on previous versions of *The Politics of Nature and*

Environmental Writing. All responsibility for any failures on my part to do so and make revisions to their satisfaction is entirely my own.

May 2025

Coburg, VIC, Australia

Competing Interests The author has no competing interests to declare that are relevant to the content of this manuscript.

Contents

About the Author

Rod Giblett is the author of 30 books of fiction and faction ('non-fiction'), including recently *Wetland Cultures: Ancient. Traditional, Contemporary* (Palgrave Macmillan, 2024) and *Waking to Nature with Thoreau and Benjamin: Or, Psychopolitical Ecology* (Bloomsbury, 2025). Forthcoming are *Wise Women for Loving Nature: Or, Femopolitical Ecology* (Bloomsbury, 2026) and *Living with Nature in the Anthropobscene: Philosophy, Ecoculture, Political Ecology* (Palgrave Macmillan, 2026). He lived by a wetland in Western Australia for 28 years, was active in its conservation for 25 years, kept a spasmodic nature journal for a few years, and published it in *Black Swan Lake: Life of a Wetland* (Intellect Books, 2013). He now lives in Melbourne, Australia, and wrote about its lost and found wetlands as a city of ghost swamps in *Modern Melbourne: City and Site of Nature and Culture* (Intellect Books, 2020). He is a graduate of the Universities of Western Australia and Sydney, and Murdoch University. He taught and researched at three universities in Western Australia for 25 years. He is the founder of wetland cultural studies, psychoanalytic ecology, conservation counter-theology, and Thoreau and Benjamin studies. He is Honorary Associate Professor of Environmental Humanities in the Writing and Literature Program at Deakin University, Australia.

CHAPTER 1

Introduction to Nature and Environmental Writing and Their Politics

Nature writing is popular. It is an enduring marketing category in the catalogues of some publishers, on the shelves of some bookshops, and in the search engines of some online booksellers, while nature everywhere is being destroyed. Is nature writing a nostalgic tourist trip, a disavowing device to ignore or overlook this destruction—the textual analogue of national parks in which special places of nature are 'set aside' for preservation while nature everywhere else is destroyed? Or is nature writing the means for nurturing bio- and psycho-symbiotic livelihoods in bioregional home habitats of the living earth, or both, or either depending on who is reading, how nature writing is read, and what is being read? I argue in this book in the affirmative for the reader, the reading, and the writing of nature writing that are engaged actively with the latter desirable option and for the reader and the reading that critiques nature and other writing that does not advocate for it.

Nature writing is often conflated or confused with natural history. Publishers, bookshops, and booksellers often lump them together. I argue in Chap. 2 that nature writing and natural history are two distinct "enunciative modalities" (to use Michel Foucault's (1970) terms). Natural history is a scientific discourse that classifies the genus and species of plants and animals in a taxonomic system. It produces abstract knowledge that exercises power over nature. It sets up a position of mastery for the subject

R. Giblett, *The Politics of Nature and Environmental Writing*, Transdisciplinary Environmental Humanities 1,
https://doi.org/10.1007/978-3-032-04427-3_1

of the naturalist and scientist to take in relation to the objects of nature. It is an institutionalized practice of seeing, stating, and meaning-making.

By contrast, nature writing engages in an intersubjective relationship in mutuality with nature. I define nature writing as the creative tracing in poetic prose or prose poetry of the bodily and sensory enjoyment of both the processes and places of nature in the broad sense of the processes and places of land, air, and water, including their plants and animals (including humans), the provision of life-giving air and energy, the temporal rhythms of the day and the seasons, and the aerial vagaries of the weather and climate. Humans are in bio-symbiosis with the oxygen-producing plants of the planet and parasitic on the planet to greater or lesser extent. Humans are also in psycho-symbiosis with the planet on a continuum from the pathological to the mutual.[1] Nature writing celebrates symbiosis and mutuality.

1.1 Nature Writing and Textual Politics

Part I of this book is devoted to a discussion of critical theory and the textual politics of nature writing, beginning with its friendly or fraught relationship to natural history. Gilbert White's *The Natural History of Selborne*, usually regarded as the first book of nature writing, was first published in 1788–1789 (White 1987). As the title suggests, White's *Selborne* is natural history in one sense as the term for him at the time in the late eighteenth century was an overarching and catch-all category for botany, zoology, geology, and palaeontology, all of which are included in his book. He was transdisciplinary before the term was invented. He also included the lists of the names of species in typical natural history style. He also advised against systematic classification as the be-all and end-all of engagement with nature. Henry David Thoreau (1817–1862) largely followed in his footsteps on all these points. Like White's *Selbourne*, a work of nature writing, such as Thoreau's *Walden*, more so his journal and many works since his time, contains some natural history.

The Natural History of Selborne is mainly a personal and autobiographical account (memoir) of White's experience of places, plants, and animals in the country outside the city. It is written in the form of letters to two

[1] For further discussion of bio- and psycho-symbiosis drawing on the work of Lynn Margulis, Margaret Mahler, and Jessica Benjamin, and on parasitism drawing on the work of Michel Serres, see Giblett (2011, chapter 12).

friends. Natural history, before, and since, is not written in epistolary form, but in scientific treatises about species of plants and animals abstracted from place. Nature writing has also been called country writing; natural history is not country writing. It can be city writing. In the nineteenth century Thoreau followed suit (without the letters) with *Walden: Or, Life in the Woods* published in 1854 (Thoreau 1982), as did Richard Jefferies with *The Story of My Heart: My Autobiography* published in 1883 (Jefferies 2002). Unlike White, both Thoreau and Jefferies critiqued modern industrial technology (as we will see in later chapters).

Unlike natural history, nature writing has not been institutionalized, despite some prominent nature writers, such as Thoreau, being canonized in American literature and sanctified in environmental philosophy principally as a nature writer of rivers, lakes, ponds, woods, and wildness. Thoreau has also been called "the patron saint of swamps" (Giblett 1996, xi, 229; discussed in Chap. 9). Swamps are a topic and topos largely outside the canon of American literature, including nature writing, and outside mainstream American culture and politics, except for the occasional reference a politician makes to draining the swamp of Washington DC, a city ironically built in a swamp according to schoolbook mythology (Giblett 2016, chapter 13). The American landscape is mountains, forests, rivers, lakes, deserts, and prairies. It is not swamps, except for the brief flourishing of a flirtation with them in the nineteenth century of which Thoreau was a pioneer (Miller 1989).

The American landscape is not marshes either, despite another prominent nature writer, Aldo Leopold, being called "the patron saint of marshes" (also discussed in Chap. 9). He has been canonized in American nature writing and sanctified in environmental philosophy principally for his land ethic, conservation aesthetic, and wilderness ethic and aesthetic. When it comes to the exclusion of swamps and marshes from the American landscape, the title and subtitle of Ann Vileisis' 1997 study, *Discovering the Unknown Landscape: A History of America's Wetlands*, says it all. Thoreau and Leopold are two founders of the American conservation movement, along with John Muir, who engaged with swamps and marshes in his twenties, but eschewed them later in his long life to become the patron saint of national parks and purveyor of the mountainous sublime (discussed in Chaps. 9 and 10). Thoreau's trajectory was from rivers, lakes, ponds, woods, mountains, and wilderness in his twenties to swamps later in his short life (discussed in Chap. 9; see also Giblett 2025, chapters 1, 2, and 9). The present-day environmental writer Rebecca Solnit critiques the politics of the American landscape (discussed in Chap. 13).

In Chap. 3 I discuss nature writing as an instance of minor literature drawing on the work of Gilles Deleuze (1925–1995) and Felix Guattari (1930–1992) on Franz Kafka (Deleuze and Guattari 1986). Rather than accepting marginalization and dismissal as "a minor genre" (Stewart 1995, xviii), nature writing wears its minority generic status as a political badge of courage and honour in the hierarchy of genres and in the struggles of genre politics that privileges the novel over every other genre. Deleuze and Guattari argue that the Czech national Kafka wrote multi-generic parafables (parables/fables, parafables, would-be fables) in the major language of German and subverted it and German imperialist culture from within. Nature writing is written in a major language, such as English, and subverts it from within. Thoreau in particular subverted mainstream American culture and the major language of American English from within written in poetic prose.

Similarly the Australian Aboriginal elder Bill Neidjie subverts mainstream Australian imperialist culture and the major language of Australian English from within in his minor literature and language of talking about what Aboriginal people call Country in the minor genre of story and mode of storytelling written in prose poetry (Neidjie 1989). Rather than nature writing, this writing is Country writing (writing about Country, Country doing the writing) in some senses similar to Gilbert White's country writing celebrating places, plants, and animals. Nature writing is also minor in the old testament biblical tradition of the minor prophets who speak truth to power, as did their major counterparts. Thoreau's and Neidjie's nature writing is politically progressive. Australian nature writing more generally, such as Eric Rolls' work, that would have benefitted from similar discussion to that of other nature writers in this book is outside its scope and word limit.

The succeeding chapters in Part I of this book draw on the cultural and critical theories of Mikhail Bakhtin (1895–1975) and Walter Benjamin (1892–1940) to read nature writing in and as the process of waking the reader to nature in psychopolitical ecology.[2] Nature writing is prosaic as it is written in prose and is about prosaic everyday life (Morson and Emerson 1990). It is also poetic in concert with the poetics of nature. It is dialogic

[2] For further discussion of Benjamin and waking to nature in psychopolitical ecology, see Giblett (2025).

between the nature writer and nature; polyphonic in a chorus of many voices, humans, and more-than-human; carnivalesque about the grotesque and the monstrous subverting hierarchies of the upper over the lower; multi-generic in memoir, excursion, essay, idyll, journal, and letter; and heterogenic in times, places, and subject matter (as discussed in the succeeding chapters). Nature writing is political in all these domains. It is also 'non-fiction,' the curious category left-over from fiction, and factual, what I prefer to call faction.

Like Kafka's multi-genericism, Thoreau's nature writing is multi-generic in memoir, excursion, essay, idyll, journal, and letter. Nature writing can be the successful, cross-fertilizing combination of these genres, or it is torn apart in the tension between them. Thoreau's *Walden* is a case in point with its essays on reading, house-building, hoeing beans, the seasons, etc.; his memoirs of his childhood and young adulthood; and his excursions in his idyllic home place. *Walden* shifts between these genres with greater or lesser degree of success or failure with combining them. Thoreau also wrote separate nature essays on various topics, such as "Walking," that differ in scope and tone from the autobiographical aspects of *Walden*. His separate essays have been collected since his death and published in collected volumes, including in *Natural History Essays* (Thoreau 1980), which would have more precisely been called 'nature writings.' The first item in this collection is in fact a review of a report of natural history, and not natural history, nor an essay, as such. His journal also contains autobiography, essays, excursions, and idylls celebrating places. His nature writing has a radical genre politics.

Aldo Leopold resolved successfully, or not, the dilemma and tension between memoir, excursion, essay, and idyll by splitting them into three separate parts of *A Sand County Almanac* with, first, his idyllic memoir of his weekend hobby and home place arranged by the months of year; second, his idyllic excursions to exotic places; and third, his "upshot" essays on concepts and themes arising from the two previous parts, such as wilderness, a land ethic, and conservation aesthetics (Leopold 1949). Along similar lines to the first, almanac part of *A Sand County*, the most recent compilation of Thoreau's nature writing on birds, and his accounts of encounters with them excerpted from his journal, assembles them into a month-by-month *A Year of Birds* (Wisener 2024).

The problem of how to integrate the genres of memoir, essay, and excursion into a successful combination is one I grappled with in *Black Swan Lake* (Giblett 2013) where I opted largely for the Leopoldian solution with a memoir in the first part arranged by months of excursions in my home place, and not to exotic places. The second part assembles "downflow" essays on topics arising from the first part, such as the seasons, the city, industrial technology, and the natural and cultural history of the black swan (discussed in Chap. 12).

1.2 Nature Writing and the Politics of Places

Part II of this book is devoted to a discussion of classic American nature writing that engages with the politics of places, such as swamps, marshes, mountains, forests, and national parks, with ways of seeing and saying, such as the sublime, and with environmental conservation. Swamps, marshes, and other wetlands have been neglected in nature writing, despite three major American nature writers, such as Thoreau, Muir and Leopold, and one minor one in Sidney Lanier, focussing on them and promoting their aesthetic and conservation values (discussed in Chap. 9). At the other extreme, forests, mountains, national parks, and the sublime have received undue and uncritical attention in the nature writing of John Muir promoting their aesthetics and nationalist values (discussed in Chap. 10). He and his successors have focussed on the conservation of special natural places (what I call sanctuarism) and neglected the conservation of nature everywhere (what is called sacrality).[3] These two chapters comprise political ecocriticism of swamps, marshes, forests, glaciers, mountains, national parks, and the sublime in the work of these four nature writers.

Rachel Carson corrected this neglect of conservation everywhere with her nature and environmental writing in *Silent Spring* first published in 1962 about the destructive impact of what she called "ecocides" on the life of plants, animals, and people in the city, suburbs, and the country. She also critiqued the militarization of American civilian life and the pollution of air, land, and water, and called for environmental conservation everywhere. Like Thoreau, Carson's nature and environmental writing was linked to progressive politics, critique of industrial capitalism, and calls for conservation as the patron saint of American conservation. She is a founder of political ecology, including feminist political ecology with her critique of masculinism (discussed in Chap. 11). Her nature and environmental writing has a radical gender politics. This book is devoted to the study of the genre, gender, and conservation politics of nature and environmental

[3] For further discussion of sanctuarism and sacrality, see Giblett (2011).

writing and to promoting nature conservation everywhere It is not a survey of the vast field of nature and environmental writing, nor a history of them, but a genealogy beginning with the birth of nature writing in Gilbert White's *Natural History of Selborne*.

1.3 Landscape Politics and the Politics of People and Places in the City and the Country

Part III of this book is devoted to a discussion of "the new nature writing" and environmental writing. During the 1990s, nature writing in the UK morphed into 'the new nature writing' and in the US into environmental writing. This contemporaneous transformation on both sides of the Atlantic was celebrated in two collections published in 2008 (Cowley 2008; McKibben 2008a). In both instances, the focus shifted from "lone enraptured males," as in the old nature writing, seeking solace and comfort in pristine places in nature ostensibly untouched by industrial capitalism and its technologies to acknowledging, confronting, and conveying the impacts and depredations of the latter on the former. Both are concerned with the more mundane, less spectacular interactions between people and places with their plants and animals in the country and the city. In the US, Thoreau writing in the nineteenth century was seen as a crucial figure in making the transition, whereas in the UK, Richard Mabey writing from the 1980s was pivotal whose writing morphed in the 1990s into "the new nature writing," a term coined by Richard Kerridge in 2001. Mabey's 'new nature writing' and that of others was included in issue 102 of *Granta: The Magazine of New Writing* (Cowley 2008) devoted to topics where very little attention was paid to the politics of people and places in the city and the country. Such was the case until recently in the 2010s in Caroline Crampton's *The Way to the Sea* and Rachel Lichtenstein's *Estuary* both about the Thames Estuary.

Richard Mabey was a leading advocate of the old nature writing by editing *The Oxford Book of Nature Writing* (Mabey 1995a) and Gilbert White's *The Natural History of Selborne* for Penguin Classics (White 1987), and writing a biography and study of White (Mabey 2006). He is also a leading practitioner of the new nature writing in such books as *The Unofficial Countryside* and *Flora Britannica* (Smith 2017), and in such essays/excursions as "The Tree of the Cross: In Search of the Fortingall Yew" (Mabey 2008; discussed in Chap. 12) published in the first foray devoted to the topic of 'the new nature writing' in a special issue of *Granta* published in 2008 (Cowley 2008). As these titles suggest, the new nature writing weaves together plants, people, and place with natural and human history in the country. The present and contemporary culture and nature are neglected though, when, for instance, Mabey does not mention in "The Tree of the Cross" that climate change is responsible for the frequency and severity of destructive hurricanes that downed twice the Selborne yew. He cannot see the wood of climate change for the yew trees.

Published with "The Tree of the Cross" in the special issue of *Granta*, "the new nature writing" of Robert Macfarlane in "Ghost Species" also weaves together plants, people, place, and natural and human history, and combines memoir, excursion, and essay (Macfarlane 2008; discussed in Chap. 12). It includes a brief autobiographical background, his account of an excursion into the Fens, and a two-paragraph summary of the history and geography of the Fens. To have gone further down the track of history and geography may have created a diversion from the main drivers of memoir and excursion of "Ghost Species." Macfarlane went into the Fens and met some local farmers who have been living there for a long time. They and some plants and animals endemic to the place are becoming extinct or "ghost species."

Yet by not delving into the history of the draining of the Fens, "Ghost Species" glosses over the dispossessing politics of drainage for the Fenlanders and the enclosing of the commons into private property, "a plain enough case of class robbery," as the late great E. P. Thompson (1968, 237) said, reproduces the mainstream agricultural narrative and paradigm, and neglects the radical, oppositional view that drainage and enclosure largely destroyed the indigenous Fenland economy and its livelihoods. The emphasis on people in the present and neglect of people in the past means that the distinctive features and history and politics of the place as a wetland commons and its drainage and enclosure fade into the background in the past and in the landscape, and become the scenic backdrop to the story about people in the present.

These two pioneering pieces "the new nature writing" of "The Tree of the Cross" and "Ghost Species" published in *Granta* forget politics. The Fens are ghost wetlands, not merely the home of ghost species, human and more-than-human. "Ghost Species" seems like a missed opportunity to consider the Fens as lost wetlands and ghost swamps in the present wetlandscape, including current projects to restore the native vegetation and cultivation of it in paludiculture (Milner and Stuart 2022).[4] Politics was not forgotten in the old nature writing of Thoreau who wrote the classics of both nature writing in *Walden* and civil disobedience in his essay on this topic. He founded political ecocriticism in his reading of selected works of the English literary canon. Raymond Williams extended political ecocriticism to the entire English literary canon in *The City and the Country* (Williams 1973). This book follows in their footsteps. With Walter Benjamin, Thoreau also co-founded psychopolitical ecology in their attention to memory, dreaming awake, and waking to nature (Giblett 2025).

Psychopolitical ecology nurtures political ecocriticism. Psychopolitical ecology diagnoses the symptoms of environ-mental illness and engages in the talking and walking cure of them in the country, especially with mourning and melancholy in relation to nature in general and wetlands in particular (as Thoreau did, and Benjamin did with the city). It also promotes environ-mental health by way of nurturing bio- and psycho-symbiosis with the living earthly household (as Thoreau also did). Psychopolitical ecology comes out of the fruitful union of the work of Thoreau and Benjamin (Giblett 2025). For them and for psychopolitical ecology the political is ecological and psychological (and vice versa). Psychopolitical ecology values materiality and spirituality equally and respects and celebrates the sacrality of the earth in the union of materiality and spirituality. It also engages with the contemporary work of Peter Sloterdijk and Byung-Chul Han within the context of the Anthropobscene, the present era of the laying bare of industrial products and wastes in the land, air, water, and the bodies of living beings and their impacts on them that were supposed to remain hidden, out of sight, and out of mind (Giblett 2026a). Nature and environmental writing brings what was meant to remain hidden out into the open and nurtures the Symbiocene, the hoped-for age superseding the Anthropobscene.

Feminist political ecology also nurtures political ecocriticism. Feminist political ecology deconstructs and decolonizes the gendered construction of reality, not only of living beings but also of processes, places and spaces, and nurtures love for them. Feminist political ecology engages with

[4] For further discussion of paludiculture in widely separated places, for example, Australia and Iraq, see Giblett (2024). For the history and politics of the draining of the Fens and the enclosure of the commons, and their mediation in art and literature, see Giblett (2024, chapter 6, 'Fen Britons').

ecofeminism and with some women thinkers and writers who are avowedly feminist, such as Rebecca Solnit, and some who were proto-ecofeminist, such as Rachel Carson. Following in Carson's footsteps, feminist political ecology critiques the militarization of life and the masculinist, misogynist, and patriarchal fascination with death that gave rise to the modern capitalist corporate-state cult of death manifested in war, especially the war against nature that includes processes, people, and places. Feminist political ecology also engages with other women writers and artists who may not self-identify as feminist whose work engages with the gender politics of nature, nurtures love for nature, and contributes to political ecocriticism (Giblett 2026b).

Nature writing has traditionally been largely concerned with the natural world as separate from the human world, or with the minor impacts and intrusion of the latter on the former, such as when the sound of the Fitchburg Railroad intruded on Thoreau's dreams and its construction interfered with and soiled the fringes of Walden Pond. The interaction between the two worlds could also be creative, such as when the construction of the railroad produced a cut near Walden Pond and created the sand foliage that Thoreau described in detail and was the upside of the intrusion (discussed in Chap. 8).

Environmental writing and the recent "new nature writing" of Caroline Crampton (2019) and Rachel Lichtenstein (2017) also foreground the interactions between the two worlds and the impacts of the human world, especially industrial technology, on the more-than-human natural world. Environmental writing for Bill McKibben is more to the point about the collision between these two worlds, a defining feature of it for him that is distinct from the old nature writing with Thoreau as a transitional figure between the old nature writing and Carson (1994) in *Silent Spring* as an exemplar of environmental writing (McKibben 2008a, xxii). Crampton and Lichtenstein trace this collision and its politics in the Thames Estuary in their books of recent "new nature writing" about it as a site of nature and culture on the margins between the city and the country (discussed in Chap. 12).

"The new nature writing" and environmental writing are often critical of the old nature writing and the old environmental conservation of special places, such as national parks, that have been aestheticized and depoliticized in landscape and wilderness painting and photography. Landscape and wilderness painting, photography, and writing celebrated the American landscape as the crucible of national identity in the bloody finale of the

Civil War and eighty years later in the threat of nuclear annihilation in the aftermath of World War II and in the subsequent Cold War. The centrality of landscape and wilderness photography and writing to American national identity is signalled and encapsulated in the National Geographic coffee-table book, *Heart of a Nation: Writers and Photographers Inspired by the American Landscape* (National Geographic 2000). There is no one unitary, quintessential American landscape, though, as the contents of the book in text and image demonstrate, except for the rhetorical, cultural, and political purposes of constructing one. This concern with landscape and national identity also gave rise to *Home Ground: Language for an American Landscape* (Lopez and Gwartney 2006), a rich lexicon of terms by sundry writers about the terms for places and landscapes.

Landscape is used uncritically in both of these books without any awareness of the history and politics of the term. It does not even rate an entry in *Home Ground*, nor does Barry Lopez discuss it in his introduction (Lopez 2006, 20–27), nor in his introduction to *Heart of a Nation* (Lopez 2000, xv–xxiv). In his introduction to both books Lopez also uses the terms "homescape," "home place," and "home ground" to describe the intimate connections and relations between people, himself included, and place. He seems to prefer these terms to landscape and shifts quickly to using them from referring early on to "the landscape where I flourish" (Lopez 2000, 20) without making an explicit case for them and this shift. "American landscape" is allowed to stand unexamined as buzzwords and a talisman for a repository of personal and national identity. American landscape becomes problematic when considering the amount of damage and destruction that the American nation has wreaked on American lands and their indigenous owners. These are, of course, not the landscapes of *the* American landscape and they have not inspired the photographs of *Heart of a Nation*, nor is the American landscape the built, urban landscapes of suburbs, cities, and towns where most Americans live (and is the heart of the nation) and photographs of them are not included either. American landscape also becomes problematic when considering suburbs, cities, and towns.

American landscape also becomes problematic when considering the history and politics of the concept of landscape. These were traced by John Barrell (1972) who argued that landscape is the view of a piece of land that was later applied to landscape painting. Barrell was following in the footsteps of the Collins pocket-size *New Gem Dictionary* (Weekley 1954, 363) that defined landscape as "a piece of inland scenery; a picture of it." Both accounts emphasize the sense of sight and the surface of the land in looking at it, and painting it. John Berger in a famous reading of Thomas Gainsborough's painting *Mr and Mrs Andrewes* critiqued the politics of landscape painting in which land is an object of sight and ownership, and an index of wealth (Berger 1972, 106–108). Jane Austen's novels exemplify these ideas of landscape, and lampoon landscape architecture and gardening to cut down trees to create pleasing prospects (Giblett 2011, 82–91). Landscape is not an object of work and a source of physical and spiritual sustenance as land was for indigenous owners and Thoreau.

'Landscape' for Bakhtin (1981, 143; his scare quotes) "is born [when] nature [is] conceived as horizon (what a man [*sic*] sees) and as the environment (the background, the setting) for a completely private, singular individual who does not interact with it." Nature "was turned into landscape" (Bakhtin 1981, 217). Landscape is born when nature dies; nature dies when nature is transformed into art (Woodring 1989; Giblett 2011, chapters 2 and 3). Similarly, contemporaneously, and simultaneously, natural history was born when nature died (as we will see in the following chapter). Landscape is an object of sight that neglects the other senses, reduces land and water to dead visual surface, ignores their living depths, and produces the aesthetics of landscape painting, and later landscape and wilderness photography (Giblett and Tolonen 2012, chapters 5–7). By contrast, Thoreau engaged with nature as inter-subject, with all his senses awake and with the depths of land and waters in ponds and swamps, and created nature writing that celebrates the life of local places, including the grotesque lower earthly stratum of swamps with their monstrous creatures, such as alligators, in psychopolitical wetlands ecology (discussed in Chaps. 7 and 9).

Against the nationalist fixation on the American landscape, Rebecca Solnit writes critically about what she calls "the landscape wars of the American west," including the dispossession of indigenous Americans for the creation of Yosemite National Park (the first in the world) and its aestheticization in the sublime photography of Carleton Watkins and Ansel Adams (Giblett and Tolonen 2012, chapter 5). Solnit also argues that "the landscape wars of the American west" later included the Nevada Test Site, what she calls "the most bombed place on earth," and its anti-aestheticization in the photography of Richard Misrach. Solnit develops the concept of "the exquisite" as the counter to the aestheticization of national parks and wilderness in sublime photography and the anti-aestheticization of feral zones and ruined places in disaster photography. Solnit follows in the footsteps of Thoreau in not forgetting politics. She is the patron saint of progressive politics and a leader in gyno- and psychopolitical ecology that embraces nature and gender and nurtures environ-mental health (discussed in Chap. 13). Her environmental writing has a radical landscape politics.

A recent instance of Canadian environmental writing and photography has a radical landscape and conservation politics. A collision between the human industrial world and the more-than-human natural world in the Anthropobscene occurred literally in 2013 when tens of thousands of migratory songbirds were incinerated in the flame of a hundred-foot-high flare at the Canaport liquefied natural gas plant in the province of New Brunswick in Canada. The LNG company concerned was convicted and fined in 2015 at a rate of less than C$100 per bird killed, an obscenely low price to put on the death of an exquisite individual living being in ornicide. The fine in total was also an obscenely low price to put on genocide, the killing of species within genuses. The story of this horrendous collision between the fossil-fuel industry of the capitalist human world and the feathered members of the more-than-human world was told nearly ten years after it occurred in text by Harry Thurstonand documented in photographs by Thaddeus Holownia in *Icarus, Birds Falling* (Thurston and Holownia 2022) . What was supposed to remain hidden, out of sight and out of mind, came into the open.

Thurston is best known as a nature writer of Canadian marshes who weaves together memoir, idyll, excursion, essay, and natural and cultural history with contemporary culture and nature.[5] He is also an environmental writer about the collision between the human industrial world and the more-than-human world with the Canaport disaster. Thurston bears witness to the disaster to it by invoking the myth of the fall of Icarus and alluding to the portrayal of his fall in a painting by Pieter Bruegel the Elder meditated upon in a poem by W. H. Auden (Giblett 2024, chapter 9). Nature and environmental writing mourn the death and celebrate the life of plants, animals, and places in the past and present and for an uncertain future in the Anthropobscene nurturing hope for living in bio- and psych-

[5] Thurston is a major nature and environmental writer whose work deserves to be, and should be, more widely known and appreciated. I have discussed his previous work extensively in *Canadian Wetlands* and recently in *Wetland Cultures*. See Giblett (2014, chapter 5; 2024, chapter 9).

symbiosis with the living earth in the Symbiocene (Giblett 2020, chapters 9 and 10).

This book is thus not a systematic survey, nor a history, of the vast and varied field of nature writing. These have been undertaken, beginning with Paul Brooks (1980), later Frank Stewart (1995), and more recently by many others, including Karen Armbruster's chapter on 'Nature Writing' (2016), Hong Cheng's *American Nature Writing* (2023), and Will Abberley's, *et al*, collection, *Modern British Nature Writing* (2022).[6] Nor is this book a genealogy of what Lawrence Buell (1995, 397–423) calls "environmental nonfiction" that he has traced and referenced extensively in Thoreau's case, nor is it a systematic survey of 'the new nature writing,' including the pioneering work of Roger Deakin (1943–2006), the late, great famous swimmer through Britain and performer of other feats of derring-do, and many successors writing in Britain, which has been undertaken by Jos Smith (2017), nor is it a systematic discussion of Thoreau's writings, nor of discussions of his writings undertaken by others, both of which I undertake to a greater extent elsewhere (Giblett 2025). Anthologies of nature and environmental writing, and the introductions to them, also survey their fields and assemble instances of both (Finch and Elder 1990; Mabey 1995b; McKibben 2008b). These surveys, anthologies, and their introductions are structured chronologically, whereas this book is structured thematically, theoretically, and politically with shifts in time and space along the way in the genealogies it traces. It also presents some long quotes from nature and environmental writing as a kind of quasi-anthology in order to illustrate and elaborate themes and theoretical concepts, express the full flavour of nature writing, preserve the integrity of the texts of nature writing, and provide exemplars for budding nature and environment writers. The creative practice of nature writing is foregrounded, but it is not 'a how-to-do-it' creative writing guide to nature writing, which has been provided by John Murray (1995), nor to environmental writing, which has been provided by Isabel Galleymore (2021).[7] It may inspire some readers to become nature and/or environmental writers.

This book is thus a celebration of the creative practices of some nature and environmental writing, a presentation of their cultural critiques, a discussion of some aspects of critical theory in relation to them, and a work

[6] I am grateful to an anonymous reviewer for drawing my attention to these three recent discussions of nature writing.

[7] I am grateful to an anonymous reviewer for drawing my attention to Galleymore's guide.

of political ecocriticism. The relationship it sets out between creative practice, cultural critique, and critical theory is at the dynamic, interactive, political, and ecological centre between them. It selects highlights and lowlights from the rich and fertile traditions of nature and environmental writing, including "the new nature writing," to illustrate and elaborate creative practice, cultural critique, and critical theory coming together in political ecocriticism. It is a thematic study, not a systematic survey, of all the features and topics of nature and environmental writing. It is the first book to do so and the first book of political ecocriticism of nature and environmental writing.

This book follows in the footsteps of Raymond Williams' pioneering work of political ecocriticism and ecocultural study of the English literary canon in *The Country and the City* (1973) that includes brief discussions of the nature writing of Gilbert White, John Clare, and Richard Jefferies. It also follows in the footsteps of Williams' pioneering cultural study of *Culture and Society 1780–1850* (1958) . Both studies do not privilege or valorize literary texts, but contextualize them within the broader culture and society, their history and politics. This book does so too as it encourages appreciative readers of nature and environmental writing to read them in new ways, and invites new readers of nature and environmental writing to appreciate their creative practice and embrace their cultural critique in gyno- and psychopolitical ecology. It encourages all readers to love and conserve the life of the earth, its species, and their habitats, hence the dedication of this book to the conservation of swallowtail butterflies (*Papilio machaon Britannicus*), their larval foodplant Milk-parsley (*Thyselium palustre*, Marsh thyme), and their only wetland home in Norfolk, England. No doubt the reader has species and habitats closer to home that require love and conservation too.

References

Abberley, Will, et al., eds. 2022. *Modern British Nature Writing, 1789–2020: Land Lines.* Cambridge: Cambridge University Press.

Armbruster, Karen. 2016. In *Nature Writing. Keywords for Environmental Studies*, ed. Joni Adamson, William Gleason, and David Pellow. New York: New York University Press.

Bakhtin, M. M. 1981. *The Dialogic Imagination: Four Essays.* Edited by Michael Holquist. Translated by Caryl Emerson and Michael Holquist. Austin, TX: University of Texas Press.

Barrell, John. 1972. *The Idea of Landscape and the Sense of Place 1730–1840: An Approach to the Poetry of John Clare*. Cambridge: Cambridge University Press.
Berger, John. 1972. *Ways of Seeing*. London: BBC/Penguin.
Brooks, Paul. 1980. *Speaking for Nature: How Literary Naturalists from Henry Thoreau to Rachel Carson Have Shaped America*. San Francisco: Sierra Club.
Buell, Lawrence. 1995. *The Environmental Imagination: Thoreau, Nature Writing, and the Formation of American Culture*. Cambridge, MA: Harvard University Press.
Carson, Rachel. 1994. *Silent Spring*. Boston: Houghton Mifflin. First published 1962.
Cheng, Hong. 2023. *American Nature Writing: Thirty Lectures*. New York: Peter Lang.
Cowley, Jason, ed. 2008. *Granta: The Magazine of New Nature Writing*, The New Nature Writing, *102*.
Crampton, Caroline. 2019. *The Way to the Sea: The Forgotten Histories of the Thames Estuary*. London: Granta.
Deleuze, Gilles, and Felix Guattari. 1986. *Kafka: Toward a Minor Literature*. Translated by Dana Polan. Minneapolis, MN: University of Minnesota Press. First published in French in 1975.
Finch, Robert, and John Elder, eds. 1990. *The Norton Book of Nature Writing*. New York: W. W. Norton.
Foucault, Michel. 1970. *The Order of Things: An Archaeology of the Human Sciences*. London: Tavistock.
Galleymore, Isabel. 2021. *Teaching Environmental Writing: Ecocritical Pedagogy and Poetics*. London: Bloomsbury.
Giblett, Rod. 1996. *Postmodern Wetlands: Culture History Ecology*. Edinburgh: Edinburgh University Press.
Giblett, Rod. 2011. *People and Places of Nature and Culture*. Bristol: Intellect Books.
Giblett, Rod. 2013. *Black Swan Lake: Life of a Wetland*. Bristol: Intellect Books.
Giblett, Rod. 2014. *Canadian Wetlands: Places and People*. Bristol: Intellect Books.
Giblett, Rod. 2016. *Cities and Wetlands: The Return of the Repressed in Nature and Culture*. Bristol: Intellect Books.
Giblett, Rod. 2020. *Odds and Ends of a Writer's Life 2014–2019*. Champaign, IL: Common Ground.
Giblett, Rod. 2024. *Wetland Cultures: Ancient, Traditional, Contemporary*. New York: Palgrave Macmillan.
Giblett, Rod. 2025. *Waking to Nature with Thoreau and Benjamin: Or, Psychopolitical Ecology*. New York: Bloomsbury.
Giblett, Rod. 2026a. *Living with Nature in the Anthropobscene: Philosophy, Ecoculture, Political Ecology*. New York: Palgrave Macmillan.

Giblett, Rod. 2026b. *Wise Women for Loving Nature: Or, Femopolitical Ecology*. New York: Bloomsbury.

Giblett, Rod, and Juha Tolonen. 2012. *Photography and Landscape*. Bristol: Intellect Books.

Jefferies, Richard. 2002. *The Story of My Heart: My Autobiography*. Totnes, Devon: Green Books. First published 1883.

Leopold, Aldo. 1949. *A Sand County Almanac and Sketches Here and There*. Oxford: Oxford University Press.

Lichtenstein, Rachel. 2017. *Estuary: Out from London to the Sea*. London: Penguin Books.

Lopez, Barry. 2000. Introduction. In *Heart of a Nation: Writers and Photographers Inspired by the American Landscape*, 20–27. Washington, DC: National Geographic.

Lopez, Barry. 2006. Introduction. In *Home Ground: Language for an American Landscape*, ed. Barry Lopez and Debra Gwartney, 20–27. San Antonio, TX: Trinity University Press.

Lopez, Barry, and Debra Gwartney, eds. 2006. *Home Ground: Language for an American Landscape*. San Antonio, TX: Trinity University Press.

Mabey, Richard, ed. 1995a. *The Oxford Book of Nature Writing*. Oxford: Oxford University Press.

Mabey, Richard. 1995b. Introduction. In *The Oxford Book of Nature Writing*, ed. Richard Mabey, vii–x. Oxford: Oxford University Press.

Mabey, Richard. 2006. *Gilbert White: A Biography of the Author of the Natural History of Selborne*. Charlottesville, VA: University of Virginia Press. First published in 1986.

Mabey, Richard. 2008. The Tree of the Cross: In Search of the Fortingall Yew. *Granta: The Magazine of New Writing 102: The New Nature Writing*, 86–96.

Macfarlane, Robert. 2008. Ghost Species. *Granta: The Magazine of New Writing 102: The New Nature Writing*, 109–128.

McKibben, Bill, ed. 2008a. *American Earth: Environmental Writing since Thoreau*. New York: Library Classics of the United States.

McKibben, Bill. 2008b. Introduction. In *American Earth: Environmental Writing since Thoreau*, xxi–xxxi. New York: Library Classics of the United States.

Miller, David. 1989. *Dark Eden: The Swamp in Nineteenth-Century American Culture*. Cambridge: Cambridge University Press.

Milner, Jim, and Judith Stuart. 2022. Paludiculture—The Future of Farming on Peat Soils? Accessed online 3rd May 2023: https://naturalengland.blog.gov.uk/2022/09/30/paludiculture-the-future-of-farming-on-peat-soils/

Morson, Gary Saul, and Caryl Emerson. 1990. *Mikhail Bakhtin: Creation of a Prosaics*. Stanford, CA: Stanford University Press.

Murray, John. 1995. *The Sierra Club Nature Writing Handbook: A Creative Guide*. San Francisco: Sierra Club.

National Geographic. 2000. *Heart of a Nation: Writers and Photographers Inspired by the American Landscape*. Washington, DC: National Geographic.

Neidjie, Bill. 1989. *Story about Feeling*. Broome: Magabala Books.

Smith, Jos. 2017. *The New Nature Writing: Rethinking the Literature of Place*. London: Bloomsbury.

Stewart, Frank. 1995. *A Natural History of Nature Writing*. Washington, DC: Island Press.

Thompson, E. P. 1968. *The Making of the English Working Class*. Harmondsworth: Penguin.

Thoreau, Henry David. 1980. *Natural History Essays*. Salt Lake City, UT: Peregrine Smith.

Thoreau, Henry David. 1982. In *The Portable Thoreau*, ed. Carl Bode. New York: Viking Penguin.

Thurston, Harry, and Thaddeus Holownia. 2022. *Icarus, Falling of Birds: A Poem by Harry Thurston, Photography by Thaddeus Holownia*. Jolicure: Anchorage Press.

Vileisis, Ann. 1997. *Discovering the Unknown Landscape: A History of America's Wetlands*. Washington: Island Press.

Weekley, Ernest. 1954. *New Gem Dictionary*. London and Glasgow: Collins.

White, Gilbert. 1987. *The Natural History of Selborne*. Edited by Richard Mabey. London: Penguin. First published 1788–1789.

Williams, Raymond. 1958. *Culture and Society 1780–1950*. Harmondsworth: Penguin.

Williams, Raymond. 1973. *The Country and the City*. London: Chatto & Windus.

Wisener, Geoff, ed. 2024. *A Year of Birds: Writings on Birds from the Journal of Henry David Thoreau*. Macon, GA: Mercer University Press.

Woodring, Carl. 1989. *Nature into Art: Cultural Transformations in Nineteenth-Century Britain*. Cambridge, MA: Harvard University Press.

PART I

Nature Writing and Textual Politics

CHAPTER 2

Nature Writing and Natural History

Nature writing is a creative practice arising out of sensual experiences of a place, including its land, water, plants, people, and other animals. Nature writing has spatial and temporal dimensions and orientations that intersect and interact in the body, mind, and writing implement of the nature writer. Nature writing defined in these terms is thus much more than simply the writing of place, or the expression of a sense of place. It is the celebration of the flows of life and energy in and between a body, a place, its plants and animals, and their natural processes. The work of Henry David Thoreau is exemplary in this regard as discussed in several chapters of this book.

I differentiate nature writing that celebrates the dynamic depths of nature from landscape writing that aestheticizes the static surfaces of nature. I define landscape writing as the creative, written inscription of the visual appreciation for the surfaces of nature in the aesthetic modes of the sublime, the picturesque, and the beautiful. The work of William Wordsworth is a prime example of landscape writing in both prose and poetry (see Giblett 2011, chapters 3 and 4). Landscape writing also features in Thoreau's work (discussed in the following chapter of this chapter; see also Giblett 2008, 58–60). Both landscape writing and nature writing can be found in the work of most nature writers (and landscape writers). Landscape writing can also be found in one anthology of nature writing with samples from such writers as Wordsworth and William Gilpin,

R. Giblett, *The Politics of Nature and Environmental Writing*, Transdisciplinary Environmental Humanities 1, https://doi.org/10.1007/978-3-032-04427-3_2

the patron saint of the picturesque (Mabey 1995a, 65–66, 69–70), but not in another (Finch and Elder 1990).

2.1 Nature Writing as Trace and Natural History as Inscription

I also differentiate between nature writing as trace from natural history as inscription. Writing as inscription on the surfaces of the body, the report, the natural history treatise, and the earth is an instrument of endo- and exo-colonization; writing as trace in the depths of the body, the slimy, and the swampy in nature writing threatens to subvert capitalist colonization and commodification. Writing is double or split between the desire to perform and to leave only traces, and the drive to inscribe and to leave durable, if not permanent, marks to erect monuments. There are not two sorts of writing, but writing is double or split between the primal writing of the trace in nature writing and the endo- and exo-colonial writing of the inscription in natural history.

This distinction between trace and inscription comes from Walter Benjamin's critical history and theory of photography from its lively birth in the daguerreotype to its commodified death in the snapshot. Benjamin regarded daguerreotypy as a trace of a thing without inscription (Benjamin 1999b, 512; see also 1979, 244). Daguerreotypy is also what Benjamin called auratic as it is a living process that traces, rather than inscribes, the unique appearance and presence of a thing, including a person, in time and space (Benjamin 1973, 222). Nature writing is also auratic as it traces, rather than inscribes, a unique appearance and presence of a living being in time and space. A nature journal with daily, dated entries made in a place traces this auratic uniqueness. Benjamin (1999b, 518; see also 1973, 222 and 224; 2002, 103–105, especially 123 n.5; 2003, 253–256) defines aura as "a strange weave of space and time: the unique appearance or semblance of distance, no matter how close it may be." Thoreau maintained perceptual distance from nature out of respect and reverence, even though he sometimes strived for greater intimacy and regretted that he could not achieve it.

Trace and aura are similar, but Benjamin (1994, 586; 2003, 106; Adorno and Benjamin 1999, 290) asserts that "the concept of the trace is defined and determined philosophically in opposition to the concept of aura." Trace is a chemical residue indicating current absence and past presence, whereas aura is a physical quality symbolizing present presence. Benjamin (1999a, 447) differentiates them in the following way:

> [T]he trace is appearance of nearness, however far removed the thing that left it behind may be. The aura is appearance of a distance, however close the thing that calls it forth. In the trace, we gain possession of the thing; in the aura it takes possession of us.[1]

Nature writing oscillates in this zone between distance and nearness, possession of a thing and possession by a thing. "What characterizes the most convincing nature writing" for Mabey (1995b, vii) "is a willingness to admit both the kindredness and the otherness of the natural world."

Consider, for example, Thoreau's account in *Walden* in which he relates that "I once had a sparrow alight upon my shoulder for a moment while I was hoeing in a village garden, and I felt that I was more distinguished by that circumstance than I should have been by any epaulet I could have worn" (Thoreau 1982, 518). In this unique moment in time, in this place in space, the sparrow takes possession of Thoreau as a perch and he takes possession of the sparrow figured as an epaulette that distinguishes the humble hoer as an officer and a gentleman. Another instance is when Thoreau wrote, "[W]hen I die you will find swamp-oak written on my heart" (cited by Richardson 1993, 17). The swamp-oak wrote on him and he wrote on the swamp-oak. He also wrote repeatedly on swamps and they wrote on him by leaving their mark on him. He is the patron saint of swamps and for psychopolitical wetlands ecology (as we will see in later chapters).

Benjamin's distinction between aura and inscription aligns largely with the distinction between strategy and tactic that Michel de Certeau (1925–1986) makes in *The Practice of Everyday Life*. Strategy is inscriptive, tactics are auratic. De Certeau (1984, xix) calls a "strategy":

> [T]he calculus of force-relationships which becomes possible when a subject of will and power (a proprietor, an enterprise, a city, a scientific institution) can be isolated from an 'environment.' A strategy assumes a place that can be circumscribed as *proper (propre)* and thus serve as the basis for generating relation with an exterior distinct from it (competitors, adversaries, 'clientèles,'

[1] For further discussion of aura, trace, and inscription drawing on Benjamin's work, see Giblett (2025, chapter 9).

> 'targets,' ['niche markets,'] or 'objects' of research). Political, economic, and scientific rationality has been constructed on this strategic model.

Natural history can be included in de Certeau's list as not only scientific rationality but also the servant of the political economic rationality of capitalism and colonialism (as we will see later in the present chapter). Again Thoreau is exemplary in this regard with his critique of capitalist political economy and its work ethic (but not of colonialism).

A strategy is a hierarchic, monologic relationship between a subject and an object. A strategy inscribes a circle between an inside proper place and an outside improper place. It surveys and patrols the frontier between civilization and the wilderness, between the clean and proper surface of the body (of the earth) and the dirty and improper depths of the body (of the earth), between the monumental upper body, the sublime upper mind and the grotesque lower body and slimy swamps, between a stable, centred sense of self and an environment out there that is removed and alien. The concept of the environment implies a separation between a subject and its environs, and a strategic relation between the former and the latter. Writing as inscription is the technological means that enables a strategic self to operate on, and in, a separate environment.

A tactic, by contrast, is a dialogic relation (to use Mikhail Bakhtin's terms) between abjects (to use Julia Kristeva's term[2]), between fluid, unstable, decentred bodies of humanity and of the earth, between self and other expressed in nature writing. De Certeau (1984, xix) calls a "tactic":

> [A] calculus which cannot count on a 'proper' (a spatial or institutional localization), nor thus on a borderline distinguishing the other as a visible totality. The place of a tactic belongs to the place of the other. A tactic insinuates itself into the other's place, fragmentarily, without taking it over in its entirety, without being able to keep it at a distance. [...] Many everyday practices (talking, reading, moving about, shopping, cooking, etc.) are tactical in character.

A tactic is auratic as it insinuates itself into the other's place, fragmentarily, without taking it over in its entirety, without being able to keep it at a distance. A tactic leaves traces of the other and the other's place. Nature

[2] Abject is Kristeva's term for that which is prior to, and refuses, the distinction between subject and object (Kristeva 1982, 1–2). The dialogic is discussed in a later chapter, and other aspects of Bakhtin's critical and cultural theory are discussed in other chapters.

writing and walking are tactical, everyday practices loved by many nature writers, such as Thoreau who walked and wrote a journal most days and wrote an essay on the topic of walking. Thoreau was inspired to walk and write in the early nineteenth century by one notable predecessor in Gilbert White and his *The Natural History of Selborne* in the late eighteenth century. Besides White, Thoreau was also writing in the tradition and context of what Lawrence Buell (1995, 397–423) calls "environmental nonfiction" whose genealogy he has traced and referenced extensively.

2.2 Nature Writing

The birth of nature writing in a single volume is usually placed with the publication of Gilbert White's *The Natural History of Selborne* in 1788–1799 (White 1987). It is written in the form of letters to two friends. Natural history, before and since, is not written in epistolary form, but in scientific treatises about species of plants and animals abstracted from place. Nature writing has also been called country writing, whereas natural history is not country writing. It can be city writing about nature in the city or the country. The locale may not matter, the species do, and the habitat may. Both Henry David Thoreau and Richard Jefferies followed suit (without the letters) in the nineteenth century with *Walden: Or, Life in the Woods* published in 1854 (Thoreau 1982) and *The Story of my Heart: My Autobiography* published in 1883 (Jefferies 2002). Unlike White, both Thoreau and Jefferies critiqued modern industrial technology (Jefferies 2002, 32; for Thoreau, see Giblett 2025, chapter 9).

White's book was one of Thoreau's favourites according to one of his biographers (Richardson 1986, 309) and the natural historian of nature writing (Stewart 1995, 27). One prominent anthology of nature writing, the monumental *Norton Book of Nature Writing*, begins with many selections from White's book (Finch and Elder 1990, 31–53). Its editors call White "the patron saint of English nature writing" (Finch and Elder 1990, 22). Another anthology, Richard Mabey's *The Oxford Book of Nature Writing*, leads up to White as a crucial break in the history of prose writing about nature, and then follows on (Mabey 1995a, 53–58). Mabey also edited White's book for Penguin Classics (White 1987) and wrote a biography of White combined with a study of his nature writing first published in 1986 (Mabey 2006). Buell's genealogy of Thoreau's "environmental nonfiction" includes White (Buell 1995, 405), whereas White for Raymond Williams (1973, 118) is significant for "the development of country writing." Recently a discussion of White's significance as "'a pioneer ecologist' whose work would lay lasting foundations for the empirical study of the natural world" leads off the editors' introduction to *Modern British Nature Writing* (Abberley et al., 2022, 1). His life and work in *The Natural History of Selborne* re-appears on many occasions in the ensuing contributions.

Thoreau's work features prominently in both anthologies. He is "more fully represented in [*The Norton Book of Nature Writing*] than any other writer" (Finch and Elder 1990, 22) and also probably in *The Oxford Book of Nature Writing* (Mabey 1995a, 79–82, 112, 120, 141–142). Bill McKibben's massive anthology of American environmental writing since Thoreau proceeds from him to Rebecca Solnit, the topic of Chap. 13 (McKibben 2008). Nature writing focusses primarily on the natural world largely to the exclusion of the human world and its impacts on the more-than-human world, whereas environmental writing for McKibben "takes as its subject the collision between people and the rest of the world" (McKibben 2008, xxii). This is no more so the case than with the death of tens of thousands of migratory songbirds incinerated in the flame of a hundred-foot-high flare at a liquefied natural gas plant in New Brunswick in Canada in 2013. This horrendous collision between the fossil-fuel industry of the human world and feathered members of the more-than-human world was documented in text and photograph by Harry Thurston and Thaddeus Holownia (2022).

Thoreau is a transitional and threshold figure between nature and environmental writing as he was for McKibben (2008, xxiv) "a tactical innovator [and] a proponent of an ethic of civil disobedience." He was also a critic of industrial capitalism, its technologies and work ethic, and of mainstream American culture, and a pioneering nature writer.[3] Thoreau for Stewart (1995, xx, xxi) is "the originator of nature writing in America" and is placed in "the forefront of nature writing" with White leading the vanguard. Thoreau for Brooks (1980, viii) is "the father of the nature essay as a literary form," though *Walden*, his most well-known book, is more than a mere collection of nature essays. It is also a memoir and critique of capitalist political economy and its technologies, such as the steam railway and electric telegraphy.[4] This combination was ground-breaking at the time and is still substantially unique in writing in English. Thoreau was

[3] For further discussion of these topics, see Giblett (2025).

[4] For further discussion of these topics too, see Giblett (2025).

the writer of nature essays, not natural history essays, despite this title being given to a collection of his essay (Thoreau 1980) and despite the title of it first piece, "The Natural History of Massachusetts," which was in fact a review of a report of natural history. It was also a manifesto for his nature writing, including foregrounding a multi-sensual appreciation for nature through the seasons (Thoreau 1980, 1–29; 1982, 31–56; 2001, 20–41).

Before White writing about nature was devoted to an animal, such as the octopus or jellyfish in Aristotle's case (Mabey 1995b, 2–3), or to other 'dumb creatures,' such as elephants and dragons in Pliny the Elder's case (Mabey 1995b, 3–5), or to denigrating woods as "wild places waste and desolate [...] of deceit and of hiding [...] of hiding and lurking" (Mabey 1995b, 11). White reversed these trends by celebrating woods as wild places fecund and fertile, and species of animals, especially birds, as sentient creatures with their own language. Landscape writing features also in White's work in which he expresses an appreciation for "commanding" from an "eminence" "the prospect" of "a pleasing park-like spot" (White 1987, 7). These were to become the clichés of English landscape design lampooned by Jane Austen in her novels (see Giblett 2011, 84–88) and critiqued by Raymond Williams in *The Country and the City* (Williams 1973, 120–126; Giblett 2012). White also appreciates "picturesque lakes and water-falls" and "lofty stupendous mountains" and voices a general preference for the beautiful over the sublime (White 1987, 99, 152). These are clichés of landscape writing and the European landscape aesthetic more generally in painting.

Rather than nature writing, or landscape writing, Williams (1973, 118) regards White as crucial for the development of what he calls 'country writing.' Rather than 'nature writing,' country writing was not only developed by White, but Country writing is also what Australian Aboriginal writers create (as we will see in the following chapter of this book). Mabey (2006, 217) wants to rescue White from the popularity of "escapist and nostalgic 'country writing'." This is not what Williams meant by 'country writing' though, as he went on to write that "anyone who lives in the country can experience at times, or seem to experience, an unmediated nature, in a direct and physical awareness of trees, birds, the moving shapes of land." Arguably there is no unmediated nature as it is always already mediated by culture, such as television and writing. Expressing "a direct and physical awareness of trees, birds, the moving shapes of land" experienced in the country and the city is the stock-in-trade of nature writing, and long may it live. Remnants of White's devotion to a direct and physical awareness of trees, birds, and the moving shapes of land live on in contention with escapism and nostalgia in country television with its programmes devoted to 'escaping to the country' permanently or temporarily.

Nature writing is the bastard child of the illicit union between literature and natural history. As a bastard, nature writing was disowned by both its parents of literature and natural history and banished to be brought up in the backblocks and the black-books, but it has been and still is immensely popular. It is a tried and true marketing category in the lists of many publishers and on the shelves of bookshops and their online simulations. White's book is "reputedly the fourth most published book in the English language" (Mabey 1987, viii; 2006, 6) after the Bible, Bunyan's *Pilgrim's Progress*, and the works of Shakespeare. Although it is the child of both literature and natural history, and so of dubious bourgeois parentage on both sides (especially when it inherits and reproduces the European landscape aesthetic), nature writing is a bastard who crosses the great divide between 'the two cultures' of the humanities and the sciences, gets shot at by both sides in the 'no-man's land' between them, and delivers a plague on both their houses (especially when it critiques both).

Despite White giving the title of *The Natural History of Selborne* to his book, it is not a natural history in several senses and ways that natural history has come to be characterized, especially in Michel Foucault's critical theory and genealogy of natural history (as we will see later in the present chapter). White was writing at a time before the descriptor 'nature writing' was invented as he effectively invented the genre. He was also writing at a time when natural history was an amateur and antiquarian pursuit and an overarching and catch-all category for any study of nature explicitly going back to Pliny the Elder in the first century of the common era whose *Natural History* was first translated into English in 1601 (Mabey 1995a, 5). Pliny's study included dragons, subscribed to the miasmatic theory of disease, and denigrated wetlands as diseased and barbarous places (Giblett 2024, 124–5). White was also an amateur student of nature across many disciplines just as they were becoming institutionalized and professionalized.

The Natural History of Selborne is also not a natural history in another sense as it is written and presented in the form of letters to two friends and is a collection in total of 110 of these letters addressed to those addressees in a conversational tone. The genre of the letter was suited to what Mabey

(1987, xv) calls White's "parochial ecology." It was also part of the trend with other literary genres of the time. The epistolary novels of Samuel Richardson were published in the 1740s and 1750s and were fashionable in the following decades in which White wrote and later collected his letters. The letter is not a genre employed in natural histories, though Thoreau's letters have been collected and published that occasionally touch on nature (Thoreau 1958, 2004). The preferred genre of natural history is a treatise with a systematic taxonomy of a genus of plants or animals, or a study of a specific species, expressed in the mode of scientific discourse, not in the conversational tone of a letter to a friend.

The Natural History of Selborne is also not a natural history in another sense as White (1987, 208) critiqued "systematic classification" and positioned it as "subservient to, not the main object of, pursuit" in botany. He advocated that the student of plants "should examine the powers and virtues of efficacious herbs, should promote their cultivation, and graft the gardener, the planter, and the husbandsman, on the phytologist," the student of the seasons. Thoreau followed suit in every respect, including grafting them together, as well as cultivating beans (as we will see in several chapters). Thoreau's nature writing was created and expressed in the genres of essay, excursion, story, idyll, memoir, and journal. White kept a journal for nearly 25 years (Mabey 2006, 146). Thoreau (1962) did so too for about the same length of time. Nature writing is a mixed genre and multi-generic. It came out of previous genres and its first writers in White and Thoreau created a new genre to express a new appreciation for nature (discussed in Chap. 6).

The Natural History of Selborne is also not a natural history in yet another sense as White emphasized in his "Advertisement" for the book that it is a book of "*parochial history*" consisting of "natural productions and occurrences as well as antiquities" (White 1987, 3; his emphases). 'Antiquities' was included in White's original title and has been dropped since without apparent warrant, though it is antiquated. The "natural productions and occurrences" included what Mabey (2006, 147) calls "the seasonal ebb and flow of natural life in a parish." Thoreau was to follow suit in spades with a sensitive appreciation for the seasons (and discussed in Chap. 6). Natural history studies the seasons in the separate branch of meteorology that also includes weather and climate. The primary focus for natural history is not natural occurrences, neither in the native habitat of their parish nor in their seasonal ebbs and flows. These processes are largely abstracted from habitat, parish, and season, and studied in isolation. The local place and its pre-history (or 'antiquities') are the background and context for these productions. Natural history can be interested in the antiquity of these productions preserved in fossils, but palaeontology is a different branch of natural history to botany and zoology. White lived in an era when they were not separate disciplines so he did not separate them. He brought all of these aspects together, including fossils (White 1987, 12), as Thoreau did in his own parochial histories consisting of natural productions and occurrences, as well as antiquities. Both were transdisciplinary before the word was invented and before the study of nature was split off into different disciplines.

The world 'parochial' comes from the word 'parish.' Mabey notes in his introduction that it

> has to do not just with geography and ecclesiastical administration, but with history and a system of loyalties. [...I]t is the indefinable territory to which we feel we belong, which we have the measure of. Its boundaries are more the limits of our intimate allegiances than lines on a map. These allegiances have always embraced wild life as well as human [life]. (Mabey 1987, xvii; see also Mabey 2006, 21)

This sense of the parish morphed into Thoreau's sense of "home is here" (discussed in the following chapter) and into the larger twentieth-century concept of the bioregion that includes the watershed and airshed of where one lives with other beings, human and more-than human, and that we call home.

White's book is also not a natural history in another sense as he did not separate himself as subject from the natural world, study it as an object, and write about it in scientific discourse free of metaphor. The nature writer, such as White and Thoreau, is a subject (or an abject) living with other living subjects (or abjects) of plants and animals in an intersubjective, or -abjective, relationship. Mabey (1987, ix) suggests that White "was perhaps the first writer to talk of animals—and particularly birds—as if they conceivably inhabited the same universe as human beings." Harry Thurston follows suit when he recounts the clash of humans with birds in the same universe in the twenty-first century and Thaddeus Holownia photographs the horrific aftermath (Thurston and Holownia 2022). White's nature writing was created by what Mabey describes as "his daily observations on weather, flowers and the comings and goings of birds"

(White 1987, xiv). Thoreau followed suit too (Allen 2019; Wisener 2016, 2024).

White practiced and promoted what he called "a more minute enquiry" into "the life and conversation of animals," including an appreciation for what he also called "the language of birds" in which "little is said, but much is understood" (White 1987, 85, 136, 152, 216). He was way ahead of his time in this regard. Again Thoreau followed suit with daily observations of the weather, plants, birds, and other animals, such as ants and squirrels (Allen 2019; Wisener 2016, 2017, 2024). White developed what Mabey (2006, 53) calls "the new notions of nature as worthy of admiration and celebration, perhaps even as a positive form or contributing partner." These new notions critiqued and refused the natural history concepts of nature as only worthy of study and taxonomization, even as a dead form, a negative force, or passive object. White expressed instead, as Mabey (2006, 96) puts it, that "the links between the human and natural world were real as well as metaphorical" and this was "to become one of the key feature of White's prose." Thoreau followed suit. Natural history largely excludes the human world. Nature writing includes the human world of people, metaphor, and other tropes. 'The new nature writing' (Cowley 2008; Smith 2017) and American environmental writing since Thoreau (McKibben 2008) do so too.

Perhaps not surprisingly there has been at least one attempt to write a natural history of nature writing, with this title. Frank Stewart's *A Natural History of Nature Writing* (1995) and Paul Brooks' *Speaking for Nature* (1980) are pioneering studies of nature writing. Both books are surveys of the field that do not come from a clearly enunciated theoretical position. They are implicitly literary histories that focus on the great authors of nature writers and the accepted canon of nature writing without an explicit theory of literature (including genre) and history (including of natural history and the human history of the rise and spread of industrial capitalism and colonialism). These are the broader contexts for the development of nature writing and its heyday in the nineteenth and twentieth centuries. The present volume engages with this context and with critical, cultural, historical, and literary theory (as we will see in the present and following chapters).

A Natural History of Nature Writing is for Stewart (1995, xx) in keeping with the attempts of natural history to construct "the taxonomy of nature writing and its possible subspecies." Thoreau is credited with creating the American genus (Stewart 1995, xx). Stewart classifies various species of nature writing and nature writer by devoting a chapter in largely chronological order to each, to eight of them, without signalling their name in the title of the chapter. He disguises the name of the author by giving each chapter the title of a quote from their writing that is a representative episode in, or prominent feature of, their nature writing. The first chapter begins with Thoreau, the second chapter goes back to his British precursor in White, and the third chapter continues by comparing Thoreau and White. Two chapters are then devoted to John Burroughs, and one each to John Muir, Aldo Leopold, Rachel Carson, Edward Abbey, and finally, and briefly, to Bill McKibben. The name of the author lurks in the background of the table of contents as the privileged writer of a representative episode or typical feature, and later re-appears in the discussion of their nature writing despite the attempt to taxonomize nature writing irrespective of who wrote it. Stewart is following implicitly in the footsteps of Brooks who is upfront about organizing his survey of American nature writing around the names of prominent authors (Brooks 1980, ix).

Michel Foucault devoted a famous essay to grappling with the question of "What is an Author?" He argued that the name of the author is privileged in modern print culture. The name of the author is also the means by which readers for Foucault fear the proliferation of meaning (Foucault 1979, 159[5]). Stewart has a bet both ways by foregrounding a representative episode or feature of the work of a nature writer in the titles of his chapter and so disguising its creation by a named author. In the second and third parts of this book I foreground nature writing's attention to a particular place or locale, such as swamps, marshes, and national park, to environmental aesthetics, such as the sublime and the exquisite, and to environmental conservation, landscapes, and their politics, and associate these with a particular nature writer who proliferates meaning. This is especially the case when it comes to swamps and marshes that have not been considered as worthy of attention in nature writing, and discussions of them, despite many prominent nature writers, such as Thoreau, Muir and Leopold, devoting attention and being attentively devoted to them (as we will see in Chap. 9).

[5] Compare an earlier abridged translation of this essay from which this argument is strangely missing (Foucault, 1977b, 138). James Birnauer and Thomas Keenan note in their bibliography of Foucault's work that the earlier translation is slightly abridged compared to the later one (Birnauer and Keenan 1987, 238).

2.3 Natural History

For nature writing to be born, natural history had to be born too and nature had to die. The death of nature for Carolyn Merchant (1980, 190 and 193) is specifically located historically in the seventeenth century, a century before the publication of White's book. For Merchant:

> [T]he new mechanical order and its associated values of power and control [...] mandate[d] the death of nature. [...] The removal of animistic, organic assumptions about the cosmos constituted the death of nature—the most far-reaching effect of the Scientific Revolution of the seventeenth century.

Merchant then goes on to discuss a number of mechanistic thinkers, including Francis Bacon and René Descartes.

The scientific revolution of the seventeenth century not only figured nature as machine but also invented natural history. The beginning of the death of nature coincided with the birth of natural history: nature as living agent and vital organism began to die, or was killed outright, when nature began to be studied as dead matter and inert object. The birth of natural history as a scientific discipline and discourse occurred in roughly 1657 according to George Seddon (1997, 10) appealing to Foucault (1970, 128), though for the latter "this date of birth is not, of course, absolutely definitive; it is there only to symbolize a landmark." Indeed, so uncertain was Foucault (1977a, 226) about locating this landmark that he later revised the date of birth backwards some 200 years by arguing that "the sciences of nature [...] were born, to some extent, at the end of the Middle Ages, from the practices of investigation." Why there was a gap of 300 years between the contemporaneous birth of natural history and the death of nature, and the birth of nature writing with White at the end of the eighteenth century is another question. Perhaps nature writing marks the end of the death of nature and the beginning of the rebirth of nature. The long period of gestation was perhaps a life-and-death struggle.

For R. G. Collingwood (1945, 5 and 95), too, writing 30 years before Foucault and Merchant "nature as machine" is an important modern invention that he associates with the so-called Renaissance and with "the denial that the world of nature is an organism," an assertion that it is "devoid of both intelligence and of life," and the development of the trope of the natural world as dead machine. The rebirth of the arts and sciences in the Renaissance coincided with, and constituted, both the death of nature as organism and the birth of mercantile capitalism.

With the rise of modernity nature was produced as an object of discourse (especially in, by, and initially for natural history) and the subject (in two senses) of science. Natural history, the first modern discourse of nature, produced nature as an object of scientific investigation and set up the subject position of the natural historian who investigated nature as object scientifically through observation. The discourse of nature produced a subject-object relationship with nature, both the first nature (and culture) of first peoples and the natural world. The development of what Foucault (1970, 157–162 and 135) calls the "discourse of nature" involved observation and representation (what he called "seeing linking spontaneously with saying"). The discourse of nature is an institutionalized way of seeing and saying the natural world and the first nature of first peoples. Nature writing, by contrast, is not institutionalized, though it has been commodified. It is not a discourse, a way of seeing and saying, but a way of sensing and writing the natural world in aura and trace.

In the discourse of nature, Foucault (1970, 160) argues, nature "is posited only through the grid of denominations without [which …] it would remain mute and invisible." Nature became visible and was made to speak in the discourse of nature through, and by, the denominative grid of taxonomy. Nature writers name plants and animals, and acknowledge "systematic classification" (as did White), but this is not the be-all and end-all, an end in-itself, but a means of engaging with a plant or animal and conveying that engagement.

The cultural relativity and severe limitations of natural history taxonomy are expressed in a purported tenth-century CE Chinese classification of the animal world in which:

> animals are divided into: (a) belonging to the Emperor, (b) embalmed, (c) tame, (d) suckling pigs, (e) sirens, (f) fabulous, (g) stray dogs, (h) included in the present classification, (i) frenzied, (j) innumerable, (k) drawn with a very fine camelhair brush, (l) *et cetera*, (m) having just broken the water pitcher, (n) that from a long way off look like flies.

This supposed excerpt from a Chinese encyclopedia was cited by Jorge Luis Borges in his essay "The Analytical Language of John Wilkins" included in his collection *Otras Inquisiciones*, translated into English in *Other inquisitions, 1937–1952* (Borges 1964, 103) and in *Selected*

Non-Fictions (Borges 2000, 231). Foucault quotes it as the virtual epigraph to *The Order of Things* (Foucault 1970, xv). Before doing so he introduces it by commenting that reading this taxonomy prompted him to laugh out loud as it shattered:

> the familiar landmarks of my thought—*our* thought, the thought that bears the stamp of our age and our geography—breaking up all the ordered surfaces and all the planes with which we are accustomed to tame the wild profusion of existing things, and continuing long afterwards to disturb and threaten with collapse our age-old distinction between the Same and the Other.

Mabey (1995b, 13) includes this Chinese taxonomy in his anthology of nature writing followed by a quote from Foucault (1970, xv):

> In the wonderment of this taxonomy, the thing we apprehend in one great leap, the thing that, by means of the fable, is demonstrated as the exotic charm of another system of thought, is the limitation of our own.

"The exotic charm of another system of thought" reduces the otherness of the Other to a curiosity in the cabinets of the museums of European colonial modernity and in the fables of far eastern orientalism (and the reaction of 'lol'), rather than acknowledging the cultural relativity and severe limitations of natural history taxonomies, let alone learning from the traditional Chinese appreciation for nature in Taoist ecology (Giblett 2023, chapter 3). With his use of parody, the fabular, and the fantastical, Borges is more attuned than Foucault to these aspects in satirizing western, 'orientalist' misapprehensions about other cultures' systems of classification.[6]

The visibility of nature for natural history was reproduced not only in the optical technologies of natural history but also in landscape, the surface of a piece of land either viewed aesthetically or painted (and later photographed) aesthetically. The sense of sight was privileged over the other senses and the ocular technologies of the microscope and telescope, and later photography and cinema, reinforced this privileging. Thoreau possessed and used a spy glass (Wood 2006, 50–52). He preferred, though,

[6] I am grateful to an anonymous reviewer for the suggestion of making more on this point.

to have "all his senses still" (Thoreau 1982, 382). His senses of sight, smell, hearing, taste, and touch engaged with nature and the sense of place.

Systematic classification made nature mute and natural history spoke for it. Nature was dragged dumb and inarticulate from what could be called the mutosphere into the semiosphere and made or taught to speak like some sort of wild child. Until the seventeenth century signs in European culture for Foucault (1970, 129) were "part of things themselves [as they still are for Aboriginal Australians as we will see in the following chapter], whereas in the seventeenth century they became modes of representation." The concept of nature from then on in European culture and its colonies represents nature (air, land, water, living beings, and so on), whereas previously the sign of nature was part of the thing itself. The former implies a highly mediated relationship between concept and referent, whereas the latter entails an immediate relationship between sign and thing. Nature writing follows suit, such as when Thoreau "nailed words to their primitive senses." Nature writing writes with nature, not about it.

The concept/referent of nature was killed in the grid of denominations and in what Foucault later called "a certain grid of explicit or implicit interrogations" just as at about the same time the grid of private property was killing the commons in the enclosure movement, the grid of drains was killing wetlands, and the grid plan town of the colonial settlement was killing Aboriginal lands (see Giblett 1996). The discourse of nature killed nature as living subject and produced nature as dead object. It also produced the natural historian as the subject with the power to exercise mastery over, and gain scientific knowledge from, and of, nature as object. What Foucault (1972, 52; see also 54–55) calls in *The Archaeology of Knowledge* "the positions of the subject" are constructed and defined by what he goes on to call "the situation that is possible for him [*sic*] to occupy in relation to the various domains or groups of objects" in what he called "enunciative modalities" (Foucault 1972, 50). These positions and situations include questioning, listening, seeing, and observing. All of these situate the subject in what Foucault (1972, 52) calls "an optimal perceptual distance" from the object. The natural historian is constructed as living subject at the expense and on the basis of nature as dead object. Nature died so that the scientist might live. Nature in this sense was short-lived; it was cut down in the bloom of its youth at fifteen hundred years old before it reached maturity and the age of majority and maturity. Nature was thereafter always going to remain in the minority. The nature writer is a living abject positioned in an intimate sensual relationship with nature as living abject. The nature writer strives for maximal sensual intimacy in close proximity with nature in all six senses, and some others besides, such as the senses of place, time, and movement (as we will see in later chapters of this book).

2.4 The Discourses of Nature

Ironically, or inevitably, the birth of the discourse of nature and the simultaneous death of nature led to the proliferation of discourses of nature. The dead body of nature was discoursed upon by long-lost relatives who had not wanted anything to do with it when it was alive. Nature is like a long-lost, wealthy aunt who dies without leaving a will and whose relatives dispute her inheritance endlessly and talk about her not having talked, or listened, to her for years. The discourse of nature—natural history—killed nature, but produced a proliferation of the discourses of nature. There is now no single and homogeneous discourse of nature, but a number of competing and heterogeneous discourses of nature.

The discourses of nature include: "the environment" that separates a subject and its environs, and produces a master-slave relationship between them;[7] the scientific objectification of nature and the subject-object relationship between them; the aestheticization of nature in landscape, especially in monumental and sublime landscapes, such as mountains, forests, geysers, and waterfalls, and in pleasing prospects and pastoral landscapes, such as gentlemen's parks and enclave estates, or either as in landscape writing; nature feminized in the virgin lands of wilderness (see Kolodny 1975 and 1984); nature misogynized in swamps and marshes (see Giblett 1996, especially chapters 2, 4, 6 and 8); nature enclosed in private property; nature preserved in sanctuaries, such as national parks and zoos; and nature commodified by industrial capitalism in producer resources and consumer "goods" and signified as an object of consumption in advertising and tourism. The body of nature was killed, but the aesthetic surfaces that represented nature were multiplied and the discursive construction of nature proliferated. Many of these discourses leaked into nature writing. For instance, the nature writing of John Muir aestheticized nature in sublime mountainous landscapes and promoted the sanctuarism of preserving special places, such as national parks (as we will see in Chap. 10).

[7] I discuss 'the environment' elsewhere; see Giblett (2008, 115–116).

Other nature writing, such as Thoreau's, refuses and resists the commodification of nature and counters with the alternative of celebrating the sacrality of nature everywhere (as we will also see in a later chapter; see also Giblett 2025).

This process of the discursive proliferation of nature was evident in the heyday of natural history in the late nineteenth century when, as Tom Griffiths (1996, 21) puts it:

> the popularization of natural history [...] was inseparable from frontier experience, intimations of war, hunting prowess, evolutionary morals, social status and manly pursuits. Natural history became an outdoor school of character formation.[8]

Natural history, not nature, became an outdoor school to which only men, especially the "coming man" of colonial modernity, were admitted: natural history was masculinized just as nature was feminized. One was not possible without the other. In fact, the two were married either de facto or legally; the masculine discourse of natural history was married in coercive control to feminized nature (see Schiebinger 1993). Yet this was also the period when the marriage was breaking down as the masculine discourse of the frontier hero in the wilderness was estranged, if not divorced, from the feminized culture of the cities. Ultimately, however, there was no conflict between the two as the masculine natural historian and wilderness hero, and nature feminized as benign and culture feminized as effete, were part and parcel of the same patriarchal paradigm (see Giblett 2011, chapter 1).

The discourses of nature police and regulate the earthly household (or attempt to) in an analogous way to the discourses of sexuality as Foucault (1981) argued in the first volume of his *History of Sexuality*. Rather than sexuality being repressed in the nineteenth century (the old orthodoxy), Foucault argued that there was a proliferation of discourses of sexuality that policed and regulated (or at least attempted to) the sexual lives of populations. Similarly, rather than nature being simply repressed or merely exploited under patriarchal capitalism, there has been a proliferation of the discourses of nature that police and regulate the lives of populations (and not just human populations). Just as the discourses of sexuality loudly proclaim that which was supposedly speechless, so the discourses of nature

[8] A similar argument has also been made by Pratt (1992, especially 15–37).

speak about the mute operations of nature. And just as the discourses of sexuality were founded on speaking the speechless, so the discourses of nature speak the mute. Everywhere nature is spoken about and for, but nowhere is nature allowed to speak for itself. Lots of talk in mainstream cultural and political discourse about "the environment," even "our environment," but little talk about our interconnectedness with the earth, about the earthly household, except in nature writing.

Within both contemporary "high" and "popular culture" nature has been overworked, represented, and commodified from theme parks to nature documentaries, from gardens to zoos, from shopping malls to ecotourism, from landscape painting to literature, from pastoralist cattle kingdoms to suburban gardens, from enclave estates to the intimate charm of the petit bourgeois lounge room with its bric-a-brac and exotic curios and souvenirs bought in and brought from other places (but often manufactured in another place). Nature is now not only a source of resource commodities but also a signifier in the service of profit. Nature, as Andrew Ross (1994, 8) has argued cogently:

> which was once assigned the lowest possible value in the account books of industrialization, is now fetching the highest market prices. Once the source simply of raw materials to be extracted and transformed into commodities, nature is now a valuable origin of exchange value in its own right. In the age of environmental accounting, nature enters the market ... for its own sake, as a desirable signifier.

"Nature" is a signifier, moreover, floating free like currency disconnected from the referent of the gold standard and within a paradigm of polysemic and multi-accentual signifieds and so a site of struggle for what it means (to use Voloshinov's terms[9]). 'Nature' has become the site of struggle between competing (class, gender, ethnic, and overlapping and intersecting ecological) interests over what and how it shall mean. Nature is not an unproblematic and universal category.

Nature nowadays is an abstract concept, collective noun, cultural construction, rhetorical figure, and sign in search of a referent. Nature can be defined as much by what it is not as by what it is as Donna Haraway (1992, 296–297) has done cogently:

[9] The sign is a site of struggle for how and what the multi-accentual signifier for Voloshinov (1973, 23) signifies in the signified.

> [N]ature is not a physical place to which one can go, nor a treasure to fence or bank, nor an essence to be saved or violated. Nature is not hidden and so does not need to be unveiled. Nature is not a text to be read in the codes of mathematics and biomedicine. It is not the 'other' who suffers origin, replenishment and service. Neither mother, nurse, nor slave, nature is not matrix, resource, or tool for the reproduction of man. Nature is, however, a *topos*, a place, in the sense of a rhetorician's place or topic for consideration of common themes; nature is, strictly, a commonplace … nature is the place to rebuild public culture. Nature is also a *trópos*, a trope. It is figure, construction, artefact, movement, displacement. Nature cannot pre-exist its construction. … Nature is a topic of public discourse on which much turns, even the earth … nature for us is *made*, as both fiction and fact … nature is made, but not entirely by humans: it is a co-construction among humans and non-humans [and the non-human].

Nature is tropical figure and topical place. Yet the topes of nature are not troped equitably. Different *topoi* are troped in different ways with some being valorized at the expense or the detriment of others. There is a politics of the tropes of topes, a politics of *trópoi* of *topois*, such as with swamps and marshes (as we will see in later chapters).

Alternatively, and even oppositionally, monstrous counter tropes deconstruct and decolonize the concept/metaphor of nature by setting up noise or interference in the communication channel about nature. For instance, Luce Irigaray (1985, 228) refers to nature as "that glass enclosure, that spangled sepulchre." Nature is here figured as fragile and transparent private property enclosed from the robust and opaque global commons and as a tomb decorated like a planetarium in which the necrophiliac (even necrophagic) rites of patriarchal capitalism are performed on the dead body of once living land and water celebrated in nature writing. Thoreau set up monstrous counter tropes in his nature writing by figuring "the most dismal swamp" as the most "holy of holies" of the inner sanctum of the temple in Jerusalem sacred to Jews and Christians alike (as discussed in Chap. 9). He subverted the conventional hierarchy that denigrated swamps and privileged temples by figuring the former as the latter.

Discourses of nature serve hegemonic class, gender, and ethnic political interests. A number of competing definitions of nature are currently in circulation, some of which see it as deadwood, as a passive object to be exploited, while others see it as a living tree, as mystical force to be revered. It is this oxymoronic figuring of nature as dead tree, as passive force, *and* as living wood, as mystical agency that characterizes hypermodernity.

Besides the discourses and tropes of nature—and growing out of them and feeding back into them—there are opposing ways of life in relation to the natural environment. Nature writing shows one way, a way of living bio- and psycho-symbiotically with the living earth.

References

Abberley, Will, et al., eds. 2022. *Modern British Nature Writing, 1789–2020: Land Lines*. Cambridge: Cambridge University Press.

Adorno, Theodor, and Walter Benjamin. 1999. *The Complete Correspondence: 1928–1940*. Edited by Henri Lonitz. Translated by Nicholas Walker. Cambridge, MA: Harvard University Press.

Allen, Francis, ed. 2019. *Thoreau's Notes on Birds of New England*. Mineola, New York: Dover.

Benjamin, Walter. 1973. *Illuminations*. Translated by Harry Zohn. London: Fontana.

Benjamin, Walter. 1979. *One-Way Street and Other Writings*. Translated by Edmund Jephcott and Kingsley Shorter. London: NLB.

Benjamin, Walter. 1994. *The Correspondence of Walter Benjamin: 1910–1940*. Edited by Gershom Scholem and Theodor Adorno. Translated by Manfred Jacobson and Evelyn Jacobson. Chicago, IL: University of Chicago Press.

Benjamin, Walter. 1999a. *The Arcades Project*. Translated by Howard Eiland. Cambridge, MA: Belknap Press of Harvard University Press.

Benjamin, Walter. 1999b. *Selected Writings, Volume 2: 1927–1934*. Edited by Michael Jennings, Howard Eiland and Gary Smith. Cambridge, MA: Belknap Press of Harvard University Press.

Benjamin, Walter. 2002. *Selected Writings, Volume 3: 1935–1938*. Edited by Howard Eiland and Michael Jennings. Cambridge, MA: Belknap Press of Harvard University Press.

Benjamin, Walter. 2003. *Selected Writings: Volume 4, 1938–1940*. Edited by Howard Eiland and Michael Jennings. Cambridge, MA: Belknap Press of Harvard University Press.

Birnauer, James, and Thomas Keenan. 1987. The Works of Michel Foucault 1954–1984. *Philosophy and Social Criticism* 12 (2–3): 230–269.

Borges, Jorge Luis. 1964. The Analytical Language of John Wilkins. In *Other Inquisitions, 1937–1952*, trans. Ruth L. C. Simms, 101–105. Austin, TX: University of Texas Press.

Borges, Jorge Luis. 2000. John Wilkins' Analytical Language. In *Selected Non-Fictions,* ed. Eliot Weinberger, trans. Eliot Weinberger, Esther Allen and Suzanne Jill Levine, 229–232. New York: Penguin.

Brooks, Paul. 1980. *Speaking for Nature: How Literary Naturalists from Henry Thoreau to Rachel Carson Have Shaped America*. San Francisco: Sierra Club.
Buell, Lawrence. 1995. *The Environmental Imagination: Thoreau, Nature Writing, and the Formation of American Culture*. Cambridge, MA: Harvard University Press.
Certeau, Michel de. 1984. *The Practice of Everyday Life*. Translated by Steven Rendall. Berkeley, CA: University of California Press.
Collingwood, R. G. 1945. *The Idea of Nature*. London: Oxford University Press.
Cowley, Jason. 2008. Editor's Letter: The New Nature Writing. *Granta: The Magazine of New Nature Writing 102*:7–12.
Finch, Robert, and John Elder, eds. 1990. *The Norton Book of Nature Writing*. New York: W. W. Norton.
Foucault, Michel. 1970. *The Order of Things: An Archaeology of the Human Sciences*. London: Tavistock.
Foucault, Michel. 1972. *The Archaeology of Knowledge*. Translated by Alan Sheridan Smith. London: Tavistock.
Foucault, Michel. 1977a. *Discipline and Punish: The Birth of the Prison*. Translated by Alan Sheridan. Harmondsworth: Penguin.
Foucault, Michel. 1977b. What is an Author? In *Language, Counter-Memory, Practice: Selected Essays and Interviews*, ed. Donald F. Bouchard, trans. Donald F. Bouchard and Sherry Simon, 113–138. Oxford: Basil Blackwell.
Foucault, Michel. 1979. What is an Author? In *Textual Strategies: Perspectives in Post Structuralist Criticism*, ed. Josue V. Harari, 141–160. London: Methuen.
Foucault, Michel. 1981. *History of Sexuality, Volume 1: An Introduction*. Translated by Robert Hurley. Harmondsworth: Penguin.
Giblett, Rod. 1996. *Postmodern Wetlands: Culture, History, Ecology*. Edinburgh: Edinburgh University Press.
Giblett, Rod. 2008. *The Body of Nature and Culture*. Basingstoke: Palgrave Macmillan.
Giblett, Rod. 2011. *People and Places of Nature and Culture*. Bristol: Intellect Books.
Giblett, Rod. 2012. Nature is Ordinary too: Raymond Williams as the Founder of Ecocultural Studies. *Cultural Studies* 26 (6): 922–933.
Giblett, Rod. 2023. *The Way of Taoism for Bodily, Spiritual and Environmental Health*. London: Transnational Press.
Giblett, Rod. 2024. *Wetland Cultures: Ancient, Traditional, Contemporary*. New York: Palgrave Macmillan.
Giblett, Rod. 2025. *Waking to Nature with Thoreau and Benjamin: Or, Psychopolitical Ecology*. New York: Bloomsbury.
Griffiths, Tom. 1996. *Hunters and Collectors: The Antiquarian Imagination in Australia*. Cambridge: Cambridge University Press.

Haraway, Donna. 1992. The Promises of Monsters: A Regenerative Politics for Inappropriate/d Others. In *Cultural Studies*, ed. Lawrence Grossberg, Cary Nelson, and Paula Treichler, 295–337. New York: Routledge.

Irigaray, Luce. 1985. *Speculum of the Other Woman*. Translated by Gillian Gill. Ithaca, NY: Cornell University Press.

Jefferies, Richard. 2002. *The Story of my Heart: My Autobiography*. Totnes: Green Books. First published in 1883.

Kolodny, Annette. 1975. *The Lay of the Land: Metaphor as Experience and History in American Life and Letters*. Chapel Hill: University of North Carolina Press.

Kolodny, Annette. 1984. *The Land Before Her: Fantasy and Experience of the American Frontiers, 1630–1860*. Chapel Hill: University of North Carolina Press.

Kristeva, Julia. 1982. *Powers of Horror: An Essay on Abjection*. Translated by Leon Roudiez. New York: Columbia University Press.

Mabey, Richard. 1987. Introduction. In *The Natural History of Selborne*, ed. Richard Mabey, vii–xxii. London: Penguin.

Mabey, Richard, ed. 1995a. *The Oxford Book of Nature Writing*. Oxford: Oxford University Press.

Mabey, Richard. 1995b. Introduction. In *The Oxford Book of Nature Writing*, ed. Richard Mabey, vii–x. Oxford: Oxford University Press.

Mabey, Richard. 2006. *Gilbert White: A Biography of the Author of the Natural History of Selborne*. Charlottesville, VA: University of Virginia Press. First published in 1986.

McKibben, Bill, ed. 2008. *American Earth: Environmental Writing since Thoreau*. New York: Library of America.

Merchant, Carolyn. 1980. *The Death of Nature: Women, Ecology and the Scientific Revolution*. San Francisco: Harper and Row.

Pratt, Mary Louise. 1992. *Imperial Eyes: Travel Writing and Transculturation*. London: Routledge.

Richardson, Jr, and D. Robert. 1986. *Henry Thoreau: A Life of the Mind*. Berkeley, CA: University of California Press.

Richardson, Jr, and Robert D. 1993. Introduction: Thoreau's Broken Task. In *Faith in a Seed*, ed. Bradley Dean, 3–17. Washington, D.C.: Island Press/ Covelo, California: Shearwater Books.

Ross, Andrew. 1994. *The Chicago Gangster Theory of Life: Nature's Debt to Society*. London: Verso.

Schiebinger, Londa. 1993. *Nature's Body: Sexual Politics and the Making of Modern Science*. London: Pandora.

Seddon, George. 1997. *Landprints: Reflections on Place and Landscape*. Cambridge: Cambridge University Press.

Smith, Jos. 2017. *The New Nature Writing: Rethinking the Literature of Place*. London: Bloomsbury.

Stewart, Frank. 1995. *A Natural History of Nature Writing*. Washington, DC: Island Press.
Thoreau, Henry David. 1958. *The Correspondence*. Edited by Walter Harding and Carl Bode. New York: New York University Press.
Thoreau, Henry David. 1962. *The Journal 1837–1861, Volumes I–XIV*. Edited by Bradford Torrey and Francis Allen. New York: Dover.
Thoreau, Henry David. 1980. *Natural History Essays*. Salt Lake City, UT: Peregrine Smith.
Thoreau, Henry David. 1982. *The Portable Thoreau*. Edited by Carl Bode. New York: Viking Penguin.
Thoreau, Henry David. 2001. *Collected Essays and Poems*. New York: Library of America.
Thoreau, Henry David. 2004. *Letters to a Spiritual Seeker*. Edited by Bradley Dean. New York: Norton.
Thurston, Harry, and Thaddeus Holownia. 2022. *Icarus, Falling of Birds: A Poem by Harry Thurston, Photography by Thaddeus Holownia*. Jolicure: Anchorage Press.
Voloshinov, V. N. 1973. *Marxism and the Philosophy of Language*. Translated by Ladislav Matejka and I. R. Titunik. New York: Seminar Press.
White, Gilbert. 1987. *The Natural History of Selborne*. Edited by Richard Mabey. London: Penguin. First published 1788–1789.
Williams, Raymond. 1973. *The Country and the City*. London: Chatto & Windus.
Wisener, Geoff, ed. 2016. *Thoreau's Wildflowers*. New Haven, CT: Yale University Press.
Wisener, Geoff, ed. 2017. *Thoreau's Animals*. New Haven, CT: Yale University Press.
Wisener, Geoff, ed. 2024. *A Year of Birds: Writings on Birds from the Journal of Henry David Thoreau*. Macon, GA: Mercer University Press.
Wood, David F. 2006. *An Observant Eye: The Thoreau Collection at the Concord Museum*. Concord, MA: Concord Museum.

CHAPTER 3

Nature Writing as Minor Literature

Franz Kafka (1883–1924) was a writer of what Gilles Deleuze and Felix Guattari (1986) call "minor literature" in a book of that subtitle. Major literature establishes the national language and reproduces its hegemony over minority languages and people. Minor literature subverts the major language from within, as Deleuze and Guattari show that the Czech writer Kafka does with German. Henry David Thoreau was a writer of minor literature who subverted mainstream American culture and the major language of American English from within (despite being canonized) in his minor literature and language of nature writing in the minor genres and non-fiction and polyphonic modes of the essay, letter, journal, nature writing, travel writing in poetic prose and prosaic poetry (as we will see with all these cases in later chapters of this book). Thoreau wrote no novel, nor short stories, and he wrote some poetry, three major genres of literature in English that privilege fiction over "non-fiction" (as we will see in Chap. 8).

Nature writing is also minor literature in the sense that its writers are minor prophets in the old testament lineage of the bible who speak truth to power. Like the book of Haggai, the books of Thoreau were produced at a time when the US' understanding of its institutions and very identity was in profound crisis triggered by slavery, fugitive slave acts, runaway slaves, and the underground railway in the preamble to the Civil War that commenced in the year before Thoreau died. Thoreau, as Haggai does, has "prophetic hope" for the "overthrow [of] the established power" of

R. Giblett, *The Politics of Nature and Environmental Writing*, Transdisciplinary Environmental Humanities 1,
https://doi.org/10.1007/978-3-032-04427-3_3

the state (Brueggemann 2019, 77). Thoreau, as Haggai does too, critiques the present political economy that "has not been [as] productive" as it might have been, or should have been because of "agricultural failure" (Brueggemann 2019, 77). Thoreau critiqued the wasteful and profligate economy and land use practices of local farmers in his psychopolitical ecology of the earthly household. Thoreau exercised justice in the prophetic sense in order to "evaluate and critique social and economic relationships," as one theologian put it in a discussion of justice (Hayes 2012, 466). Haggai, like Thoreau too, is devoted to "cleanness, purity and holiness" (Brueggemann 2019, 77). Thoreau was an ascetic and pre-secular holy man who followed in the footsteps of yogis and their laws. Following also in the footsteps of the final three minor prophets, Thoreau's objective was to encourage spiritual seekers to live a simple life. Like Haggai, Thoreau's literary style is simple, direct, and festive. Or it can be pompous, prudish, and condescending.[1]

Contrary to popular opinion, rather than predictors of the future, the biblical prophets were primarily what Walter Brueggemann (2019, vii; his emphases) calls emphatically "**emancipated imaginers of alternative**." Similarly, for Heschel in his classic, monumental monograph, *The Prophets*, "the prophet is a poet. [...] Like a poet, [s/]he is endowed with sensibility [...] and above all, with a way of thinking imaginatively. Prophecy is the product of poetic imagination. *Prophecy is poetry*" (Heschel 2001, 468–469; his emphases). Along similar lines to Heschel, Brueggemann (2018, 45–46) in his classic monograph called the prophet's sensibility and way of thinking "the prophetic imagination." He also suggested that the "poetic imagination is the last way left in which to challenge and conflict the dominant reality" (Brueggemann 2018, 40).

The poetic imagination and the prophetic imagination work in concert with two other sensibilities and ways of thinking, two other "imaginations," or aspects of the imagination, pertinent culturally and politically to nature writing, especially to the activity of writing and the use of language. The first is what Lawrence Buell (1995) calls "the environmental imagination" expressed in nature writing whose exemplar and pioneer for him is Thoreau who wrote in poetry, prosaic prose, and poetic prose about nature. The other sensibility and way of thinking pertinent to nature

[1] For further discussion of the minor biblical prophets and the major nature writers (Thoreau, Muir, Leopold, and Carson) and other environmental apostles, see Giblett (2021a).

writing is what Mikhail Bakhtin (1981) calls "the dialogic imagination" (discussed in a later chapter).

The Australian Aboriginal elder Bill Neidjie was also a writer of minor literature and a minor prophet with a poetic, prophetic, environmental, and dialogic imagination who created poetic prose. He was born in 1911, 1912, or 1913 along the East Alligator River in the Kakadu region of the Northern Territory in Australia and passed away in 2002. There are no alligators in this river, but there are saltwater crocodiles. The mistake made by Captain Phillip Parker King in naming the river continues to be made in an ongoing act of colonization. Neidjie told what he called his "story about feeling" in English to Keith Taylor in 1982 (Neidjie 1989). He subverted mainstream Australian culture and the major language of Australian English from within in his minor literature and language of talking about Country in the minor genre of story and in the mode of storytelling. Neidjie's story is told in prose and presented on the printed page as poetry as the line breaks give pause for contemplation and meditation. *Story About Feeling* is prose poetry. Thoreau, Neidjie, minor literature, Country, nature writing, and storytelling are the intimately intertwined topics of this chapter.

3.1 Writing a Word for Wetlands

Neidjie and Thoreau also wrote a word for wetlands in wetland literature that is doubly marginalized within the marginalized literature of "nature writing" in psychopolitical ecology. Wetland literature has been marginalized in the canons of major national literatures, even within the opus of canonical writers, such as Thoreau and his writings on swamps and Aldo Leopold and his writings on marshes, both of which were neglected or overlooked for a long time. Wetland literature is doubly marginalized within the marginalized literature of "nature writing" and "country writing," and in critical writing about them, for example, John Clare's writing about the Fens and Raymond Williams' writing about it (both discussed later). This radical political slant on the context for wetland conservation has not been addressed in global policy on wetlands.

Marginalized communities are usually regarded as human communities composed of gender and sexually diverse, gay, lesbian, bisexual, asexual, queer, non-binary, intersex, and transgender + peoples forming the rainbow coalition. Wetlands are also non-binary: neither land, nor water; neither solid, nor liquid. Their identity is fluid, trans-generic, and indeterminate. Wetlands are queer bodies of water. This radical political slant on the context for wetland conservation within psychopolitical wetlands ecology has also not been addressed in global policy on wetlands. People of colour are also a marginalized community. Wetlands are places of colour and marginalized communities of plants and animals. More specifically, wetlands in mainstream western culture and its colour-coding are black with all its incipient racist associations.[2] Wetland figured as "black waters" would be a convenient shorthand label to encapsulate more than two millennia of mainstream western denigration (literally "blackened") and destruction of wetlands (see Giblett 2021b). Wetlands are black waters, distinguished from white water and discriminated against accordingly in word and deed by whites (see Giblett 2007, 2009, chapter 11).

Marginalized communities of peoples act and agitate in solidarity with each other and their allies in common cause—social, cultural, and political. As such, they are anthropocentric and do not consider and act in solidarity with other diverse, marginalized communities, such as wetlands with their plants and human and more-than-human animals. Wetland communities could be a part of a much broader, more inclusive rainbow coalition that values the green as much as the other colours in psychopolitical ecology. I am yet to find a section in any bookshop devoted to wetland literature that sits side-by-side with the Aboriginal, indigenous, gay, lesbian, queer, bisexual, non-binary, and transgender + sections.

Wetlands are marginalized communities partly because of their identity. Wetlands have been gendered, or more precisely feminized, in the gendered construction of reality. Wetlands are feminized for either good or ill: ill in misogynist patriarchy as places of disease, death, horror, the monstrous, and tombs of dead matter; good in matrifocal ecofeminism, especially French feminism, as places of life, death, health, the maternal, and wombs of new life.[3] The latter deconstructs the binaries of life and death, womb and tomb, monstrous and maternal. Wetlands are places of both gyno- and psychopolitical ecology.

Dredging, draining, and reclaiming wetlands is a quintessentially masculinist thing to do. It was certainly the kind of war the fascist Benito Mussolini liked to fight and win in the 1930s. The authoritarian Saddam Hussein did likewise with the marshlands of southern Iraq in the 1990s (Giblett 2024, chapter 3). Floods in wetland cities, such as Venice, New

[2] For further discussion of black, see Giblett (2013, chapter 14).

[3] For further discussion of all these aspects, see Giblett (1996).

Orleans, and New York, suggest that in the longer term the war was never won, never could be and at best only a truce could be achieved in the short term with the will to fill or drain wetlands and under the drive to dredge, drain, and reclaim.

Part of the problem with wetlands for mainstream western culture lies in their aesthetic qualities, or lack of them. Wetlands often look ugly, smell bad, feel slimy, sound eerie, and are uncanny. As a result, they have been denigrated, or they have been aestheticized occasionally within the conventions of European landscape aesthetics in the drive to fill their lack of aesthetic qualities and remedy their sensory ills.[4] This aestheticization occurs not only in landscape painting and writing but also in landscape architecture by transforming wetlands into lakes in parks and gardens.[5] By contrast, Aboriginal wetland peoples have used and appreciated wetlands for tens of thousands of years, including in their art and literature.

3.2 Aboriginal Country

Traditionally for Australian Aboriginal people loving nature, including living beings and places, means loving Country and being loved by Country. Australian Aboriginal people do not create nature writing, but Country writing. They largely write in English and create a minor literature that disrupts and subverts the major language of English from within. Kafka was a colonized internal exile, not only because he was a Czech national in the German-speaking Austro-Hungarian empire but also because he was Jewish by birth. Aboriginal people are colonized internal exiles within the British and Australian empires with their own religion of sacred beings, such as the Rainbow Serpent. Gershom Scholem (2012, 190–191) argued that Kafka, even while being "tied to the German language and its intellectual world, never succumbed to the illusion of being at home" in it because he was not of this world at all. Most Aboriginal people are tied to the English language by necessity, but not to its intellectual world and have never succumbed to the illusion of being at home in it because their home is in Country, even though their own language may be lost.

[4] For further discussion of aesthetics and wetlands, including their uncanniness, see Giblett (1996, part I), and wetlands in Australian (including Aboriginal) painting and photography, see Giblett (2020a). For the uncanny in philosophy, psychoanalysis, culture, nature, cities, technology, and literature, see Giblett (2019).

[5] Many instances spring to mind, such as Central Park in New York (see Giblett 2016, 149–152) and the Royal Botanic Gardens in Melbourne (see Giblett 2020b, chapter 7).

Country is what Marcia Langton (2000) calls "homeland." Country is the bond between a people and their place. Country is homeland, a biogeographical and linguistic place of being and belonging, of dreaming, songs and stories of a watershed, airshed, foodshed, and geological formations shared with endemic native plants and animals who make it their homeland too and make it into farms and gardens without fences, compared to farms and gardens with fences, the conventional stock-in-trade of Anglophone nature writing.

Country is thus cultural, not natural. Country is cultural landscapes, "not natural landscapes," as Michael-Shawn Fletcher puts it (Cumpston et al. 2022, 68). For Mick Dodson:

> [C]ulture transforms land and environment into landscape, and into 'country'. ... Culture is what enables us to conceive of land and environment in terms that are different to conventional European notions ... 'landscape' and 'environment' are human constructs. (cited by Langton 1998, 7)

Landscape is a problematic term. It came into English as a pictorial representation of the countryside in painting. It is land laid out and presented as scenery. It was later applied to the countryside in general, and still later to a place (Barrell 1972, 1). Nowadays scientists use it as a catch-all category for landforms and biochemical processes. Others use it to refer to a piece of land, or to the land in general, or to sound sophisticated by using a longer and abstract word when the shorter and concrete "land" would do and say what they mean. More specifically and historically, landscape refers to the surface of the land. It denies and represses the depths of the earth and water that give land life, such as wetlands. It privileges the sense of sight and observing land from a single point of view. It reduces land to a visual phenomenon and ignores the other senses that engage with land and the earth. It composes the land into pictures ready to be sketched and painted, or to be viewed from a stopped car, or captured on camera.

By focussing on the surface of the land, landscapes are feminized in the European aesthetic conventions of the beautiful and the picturesque with their focus on smooth surfaces, whereas landscapes are masculinized in the European aesthetic conventions of the sublime with its obsession with the heights of rugged mountains, and other massive and monumental structures. Thoreau feminized the surface of the earth in a picturesque description of a landscape in his journal. Thoreau's representation of the landscape is as an imagined or remembered, or strictly phantasized and feminized body, the phantasmatic body of the mother with its "undulations," "its nerves and muscles" (Thoreau 1962, *XII*, 74–76). The sight of the land evokes the touch of a body "firm and soft [almost a contradiction in western terms] as velvet." Yet the earth is still a body viewed from afar, a body of smooth surface, a screen, onto which the Thoreauvian viewer projects his phantasies of a full and replete body. Landscape enables the viewer to remain distant and aloof. The closest he comes to intimacy is through memory and imagination. The landscape remains removed; the land is reduced to phantasized body. Thoreau wrote later about swamps in which he upset European aesthetic conventions and appreciated their life-giving depths (as we will see in Chap. 9).

The entire country of Australia (dry land as well as wetlands, rivers and seas—the land under the water) was, and still is, Aboriginal Country that precedes and refuses the modern European distinctions between the city and the country, culture and nature, cultural and natural landscapes, park and wilderness. Australian cities and suburbs are built on Aboriginal Country, not only Aboriginal in the sense that the land was owned and worked by Aboriginal peoples, but also the original land itself. Aboriginal Country is the country of Aboriginal people who live not only in the country but also in the city and the towns. Country is in the city and the country. Country is not only outside the city, but also inside, beneath, and before the city. The city has not expunged Country. "All places in Australia, whether urban or remote, are Country," as Zena Cumpston (2022, 12, 33) puts it. All people in Australia live on Aboriginal Country. All of Australia is Aboriginal Country. This tradition is carried on in contemporary times in caring for Country, and in acknowledging country and welcoming visitors to Country in all parts and places of Australia.

All of Australia is Aboriginal Country, but parts of it, such as agricultural, pastoral, and mining country, and the cities and suburbs, are also European landscapes laid over and constructed on top of Aboriginal Country. There are no areas or parts of Australia, including under the sea (there is sea-Country and deep-sea dreaming) and in the sky (there is sky Country and dreaming astronomy), that are uncreated by ancestral beings or untouched by human hands. Langton (1998, 14) describes how, "in the Aboriginal cosmology, humans and non-humans are related in special ways, as if they were kin, through their common descent from the ancestral beings." Country is kin and kindred.

Country is thus not wilderness. Deborah Rose (1988, 384 and 385) puts it bluntly that "in Australia, there is no wilderness […] wilderness is actually home." Langton (1998, 18) states emphatically "there is no wilderness, but there are cultural landscapes." There are no non-cultural landscapes in Australia. Aboriginal Country is not only a cultural landscape and the country outside the city but also the city, the land, or country where the cities and the suburbs now stand, the alien European cultural landscape imposed on the land. There are also no areas or parts of Australia that are untouched by modern industrial technologies, including satellite surveillance and photography.

With Aboriginal Country there is no landscape to be culturally constructed or greedily consumed, no discourse to produce an object and a subject, but a loving union of people and place, culture and Country. Aboriginal Country is not a construction, so it cannot be deconstructed. It is not a discourse, an institutionalized way of seeing, saying, and doing, but a symbiosis, an intimate and intuitive way of being and living. "Country is physical, communal and spiritual—land, water, sky, habitats, sites, places, totems and relationships, a world of the mind, a way of being and behaving," as Bill Gammage puts it in *Country* (Gammage and Pascoe 2021, 81–82). Country does not pursue a drive for mastery and possession, but seeks a desire for mutuality and reciprocity. Aboriginal Country is outside history (white chronological history); it is (in) pre-history; it is (in) the Dreaming; it is not in the past, but in the past, present and future, the now, the now-time. It precedes and refuses the colonization of time in chronology. It is what could be called a "natural cultural landscape" to oppose the "cultured natural landscapes" of the gentleman's park and enclave estate, national park, wilderness, mining, pastoralism, the bush, and the city (see Giblett 2011 for further discussion of these topics).

There are no parts or places of Australia that are unloved. Country is loved and loving. *Loving Country* is the tile of a recent "guide to sacred Australia," aka Aboriginal Country, by Bruce Pascoe and Vicky Shukuroglou (2020). It is a guide for travellers to and in Australia, including armchair travellers, aka readers, to places and sites of significance for Aboriginal people around Australia. Loving Country involves Country loving and being loved. Country is loving, and Country should be loved in return. The guide invites travellers to love Country, to treat it and its places with respect. It is also an invitation to participate in Aboriginal spirituality and create a better, more loving, less hateful, Australia. Pascoe and Shukuroglou (2020, vii) "believe that the better angels of Australia want a more […] loving Country. That lovingness may decide on a treaty between all Australians." Aboriginal lovingness "sees women as the centre of civilization," as Pascoe puts it (Pascoe and Shukuroglou 2020, 9). This guide to loving Country (in two senses) fills a gap left by "the failure of Australian tourism to tell the Aboriginal story" (Pascoe and Shukuroglou 2020, 302). It provides a template for Australian tourism to do so and a resource for tourists and travellers (visitors and local alike) in Australia to hear the Aboriginal story. It could lead to all people loving Country.

3.3 Bill Neidjie and Kakadu Dreaming

Aboriginal writing about Country embraces the conventional elements and topics of Anglophone nature writing, such as plants and animals, rocks and water. This is the case, and much, much more besides, with Bill Neidjie's *Story About Feeling* about living in Kakadu in northern Australia before and after it became a national park, now "Australia's largest terrestrial national park" (Langton 2021, 246). Marcia Langton (2021, 246) points out that "Aboriginal people have occupied Kakadu for 65,000 years, the longest record of continual occupation of any area on earth." Aboriginal writing about Country also includes the heavens and the stars, often not discussed in Anglophone nature writing with the notable exception of Richard Jefferies (2001, 17–21[6]), but surely a part of nature. Bill Neidjie in *Story About Feeling* connects the heavens and stars above to the earth and everything below:

> Because all that story up there
> and here, all this dream here
> that's the one up there.
> Star up there but here e made something…
> rock or billabong or might be river.
> That way star there. (Neidjie 1989, 10)

"Star up there" includes the living crocodile formation floating in the Orion constellation in the Milky Way (Neidjie 1989, 5, 135). In the glossary to *Story About Feeling* "e" is "he, she or it" (Neidjie 1989, 176). "E" is of indeterminate gender. "E" is transgender and queer. "E" is not

[6] For further discussion of the neglected nature writing of Jefferies and his writing a word for wetlands, see Giblett (2021a, chapter 32).

inanimate, but a living being, not dead matter. "E" made the living water of the billabong, or "flood plain." Neidjie (1989, 157) says that "flood plain they call it now." Flood plain implies an aberrant or abnormal influx of water, whereas billabongs and other Australian wetlands have an inflow and outflow of water that is normal and typical. Thoreau had a similar view of rivers and meadows (as we will see in a later chapter). Meadows and billabongs are queer bodies of water. Their water "regime" is highly productive. Neidjie (1989, 157) says "your food in all that water [...] good to eat in that billabong" (Neidjie 1989, 157). In general, Neidjie (1989, 167) says "nice country in the plain."

For Neidjie (1989, 44) "our culture, our tucker" includes yams and the injunction that when "you dig yam":

> [Y]ou digging through the belly.
> You must cover im up, cover again.
> When you get yam you cover
> so no hole through there
> because you killing yam other thing
> And you got to hang on. (Neidjie 1989, 25)

This injunction is a specific aspect of the general injunction that "this earth, this ground, this piece of ground e grow you" (Neidjie 1989, 30). Neidjie (1989, 150) reiterates later that "e give us something to eat, give us life/tree, water, grass, yam." In general, "this earth e grow, you growing little by little" (Neidjie 1989, 166). Earth grows and grows people in symbiosis.

E grows culture and body. Neidjie said:

> This culture belong to us.
> In our body with this, this story. (1989, 103)

This story of and about feeling is told in words and painting (Neidjie 1989, 113). It is to be felt, not read in print, or seen in paint. Story is felt in the body. Neidjie said:

> [W]e like story
> because story going with your body [...]
> I feel in my feeling with this story. (1989, 121)

Feeling connects people and country. Neidjie says he is:

> Feeling with my blood or body,
> feeling all this tree and country [...]
> E pumping now that leaf, that blood there.
> E not wood but you'n'me that [...]
> feeling make you.
> Feeling make you there with wind, open place
> because e coming through your body [...]
> In my body that tree. (1989, 167, 168, 170)

All animals, including humans, are in symbiosis with the oxygen-producing plants of the planet. Neidjie also felt that tree in his body with the same blood flowing in both.

Neidjie tells the "White-European" who asks "What for?" to "listen carefully":

> Because e might feel his body with the blood, water,
> feeling. Story good for you.
> That way e build this world for us.
> That way that story ...
> We came this world, animal and man and woman. (1989, 121)

Through and with story white Europeans might feel their body with its blood and water continuous with the water as the life-blood of the body of the earth, as one body. Story builds the world, one world.

Neidjie was a critic of the Ranger Uranium Mine in Kakadu National Park (see map Neidjie 1989, 174–175). Aboriginal people were not consulted about its development in a shameful betrayal of Aboriginal land ownership and of social and environmental justice. Neidjie (1989, 148) said: "[W]e might get sick [...] because uranium is poison." Mining commenced in 1981, ceased in 2012, and processing was completed in 2021. Rehabilitation is ongoing, expensive, and contentious. There were many breaches of the environmental conditions placed on the mine, including, alarmingly, with millions of litres of radioactive water flowing into Kakadu National Park in April 2010 as reported under the headline "Kakadu being Poisoned" (Murdoch 2010). This was not a one-off event, but is an ongoing process. Poisoning country colonized time as well as space with the half-life of radioactivity. Poisoning country poisons people, place, and time. Neidjie said:

> If somebody take im your country, you'n'me get sick.
> Because feeling … this country where you brought up
> and just like you'n'me mother. (Neidjie 1989, 153)

For Neidjie (1989, 170) "earth is my mother or my father." Who would poison their mother or father? Some did.

Neidjie's *Story About Feeling* was told to, and recorded and transcribed by Keith Taylor, and is presented as poetry. It is the transcription and presentation of an oral story. It disrupts and subverts not only the major language of English and the conventions of Anglophone nature writing from within, but also the major European genres of story and poetry by creating what Gary Saul Morson following Bakhtin calls a prosaics of everyday life (discussed in the following chapter). It also disrupts and subverts the experience of being a colonized internal exile by finding home in Country and locating "that tree" "in my body."

3.4 Storytelling

Storytelling for Walter Benjamin (1973, 91; 2002, 149) in his essay "The Storyteller" is "itself an artisanal form of communication. […T]races of the storyteller cling to the story the way the handprints of the potter cling to a clay vessel." Traces of Neidjie cling to his story in its published form with a photograph of him, Keith Taylor's brief biography, and his creative art of English-speaking (Neidjie 1989). Traces of Thoreau the storyteller cling to his stories with the movements of his writing hand in his handwriting as he wrote with a quill pen, the last one of which is housed in the Thoreau Collection at the Concord Museum (Wood 2006, 137). Neither Neidjie nor Thoreau explains their stories. For Benjamin "it is half the art of storytelling to keep a story free from explanation as one recounts it" (Benjamin 1973, 89; 2002, 148). This is certainly the case with Thoreau's famous cock and bull story in *Walden* about "a hound, a bay horse and a turtledove" (Thoreau 1982, 272; discussed in Chap. 8). To explain a story, is to reduce it to information. Information for Benjamin (1973, 89; 2002, 147) is "incompatible with the spirit of storytelling." Storytelling is engaging and embodying, whereas information is alluring and seductive. Benjamin (1999, 72) describes how "the youthful goddess of information performs her provocative belly dance," whereas the elderly goddess of storytelling weaves her inviting tapestry to walk into and with. Storytelling engages the listener from a decent and mutually respectful distance delivering the story discreetly, whereas information tantalizes the informant with the seductive temptation of closeness and intimacy. When Thoreau is providing information about plants, ponds, or birds, he is not telling a story with spirit. On many other occasions, he does tell the stories of plants, ponds, and birds, with spirit. Neidjie never provides mere information about plants and animals and his stories are told with spirit.

Both Thoreau and Neidjie did not write novels. The novel for Benjamin is not storytelling. He argued that "the decline of storytelling is the rise of the novel" (Benjamin 1973, 87; 2002, 146). This rise and fall are inextricably linked in a countervailing motion like a see-saw. The decline of storytelling with the rise of the novel is not only due to the development of the printing press without which the novel is impossible. Nor is it only due to the development of mass literacy without which the novel is impossible too. It is partly due to the development of the reader usually reading alone. Writers usually write alone too. This is the novel-writing and -reading social contract. Benjamin argues in the same essay that "the birthplace of the novel is the individual in his isolation" (Benjamin 1973, 87; 2002, 146). This applies equally to the novel reader and the novel writer. By contrast, the birthplace of the story is a group of listeners gathered together in an assembly of auditors to hear the story. Neidjie's *Story About Feeling* invokes and replicates this setting and sensation of being among other listeners listening to his story. For Thoreau the storyteller in his isolation in Walden Woods the challenge was to use print as a means to assemble a readership and to speak directly to it. Like the novelist, he was an isolated writer writing to an isolated reader. Unlike the novelist, he wrote in and about the company of birds, squirrels, ponds, swamps, and woods. For Neidjie the storyteller in company with Keith Taylor in his Country of Kakadu with its plants, animals, and waters like Thoreau the challenge was to use print as a means to assemble a readership and to speak directly to it. Both Thoreau and Neidjie met this challenge.

3.5 Home Is Here

Thoreau is also a similar figure to Neidjie for finding his home here and being at home in his own body and home place, and for writing about both. Writing in his journal four days after his thirty-fourth birthday he recalled "in youth, before I lost any of my senses, I can remember that I was all alive, and inhabited my own body with inexpressible satisfaction; both its weariness and its refreshment were sweet to me" (Thoreau 1962,

II, 306–307). Thoreau looks back with a tinge of longing and regret for the bodily presence of mind he had in his youth. Nearly nine years earlier when he was twenty-four years old he had already expressed an exilic experience of his own body when he wrote in his journal that "I must confess there is nothing so strange to me as my own body. I love any other piece of nature, almost, better" (Thoreau 1962, *I,* 321).

Thoreau not only looked back to the past to the home of his own body but also looked in the present to home in places and the future of places. In his youth he was at home in his own body, whereas in his maturity he was not as at home in his body as he was in his youth, but he was more at home in his own country, his homeland. As he grew up and grew older his home changed from his body to his places. Many years later he wrote in his journal that "home is here" (Thoreau 1962, *V,* 254). Several years later he elaborated in his journal:

> Here I am at home. In the bare and bleached crust of the earth I recognize my friend. [...T]he constant endeavour should be to get nearer and nearer *here.* [...] A man dwells in his native valley like a corolla in its calyx, like an acorn in its cup. *Here,* of course, is all that you love, all that you expect, all that you are. Here is your bride elect, as close to you as she can be got. Here is all the best and all the worst you can imagine. (Thoreau 1962, *XI,* 275; his emphases)

Home is here; it is not there, some place elsewhere. Thoreau wrote earlier in his journal that "the man who is often thinking that it is better to be somewhere else than where he is excommunicates himself" (Thoreau 1962, *X,* 190–191). "He" is no longer in communication with the sacred place of home, but has banished and excommunicated "himself" into exile, whereas Thoreau communicated himself into a non-exilic experience of being at home in his home place.

Along similar lines, Richard Jefferies describes how:

> Mechanism increases convenience—in no degree does it confer physical or moral perfection. [...] Control of iron and steel has not altered or improved the bodily man. [...] Our bodies ae now conveyed all [a]round the world with ease, but obtain no advantage. As they start so they return. The most perfect human families of ancient times were almost stationary. (2002, 32)

The conveying of bodies around the world recently obtained a decided disadvantage in the spread of the Corona virus and conferred physical imperfection, unimprovement, and death.

Jefferies was not only a critic of modernity, but also a prescriber of the antidote to its ills that he conveyed in his nature writing. He did not simply prescribe escaping to the country as a pastoral retreat from the ills and evils of the city, but nurtured new ways of being and living with the earth in the here and now (as we will see in Chap. 5).

Thoreau's inhabited not only his own body in his youth but also his own home of his country in his maturity. In his first book, *A Week on the Concord and Merrimack Rivers*, he enjoined his readers: "leave not your native land behind" (Thoreau 1998, 116). He wrote in his journal that "this is my home, my native soil. [...] Here have I my habitat" (Thoreau 1962, *III*, 95). His native land and soil, his habitat, were not only his own body, but also what is now called his bioregion (see Giblett 2011, 242–246). Thoreau enjoyed living and walking on the urban fringe in the swamps and woods and rowing and sailing on the rivers around Concord in Massachusetts. Thoreau also pointed to the larger home of the earth and the world. When he was twenty years old he wrote in his journal: "I am at home in the world" (Thoreau 1962, *I*, 11). The world for the young Thoreau was a homely place. The older Thoreau was well aware that it was not always so. When he was thirty-three years old he wrote in his journal: "[C]ommonly I am not at home in the world" (Thoreau 1962, *II*, 76). When he was forty-two years old he asked in a letter: "[W]hat is the use of a house if you haven't got a tolerable planet to put it on?" (Thoreau 1958, 579; 2004, 181). This question is even more urgent in the age of global boiling making the planet intolerable and destroying houses with feral fires, floods, and landslides. Without a house and being homeless, the world is an unhomely place. The antidote and cure is living simply and sustainably in the home place of a bioregion, being engaged in its history, being embodied sensorily with the life of its plants and animals, being active in the politics to conserve or rehabilitate it, and living lovingly on the living earth. Nature writing shows ways of doing so.

References

Bakhtin, M. M. 1981. *The Dialogic Imagination: Four Essays.* Edited by Michael Holquist. Translated by Caryl Emerson and Michael Holquist. Austin, TX: University of Texas Press.

Barrell, John. 1972. *The Idea of Landscape and the Sense of Place 1730–1840: An Approach to the Poetry of John Clare.* Cambridge: Cambridge University Press.

Benjamin, Walter. 1973. *Illuminations.* Translated by Harry Zohn. London: Fontana.

Benjamin, Walter. 1999. *Selected Writings, Volume 2: 1927–1934.* Edited by Michael Jennings, Howard Eiland and Gary Smith. Cambridge, MA: Belknap Press of Harvard University Press.

Benjamin, Walter. 2002. *Selected Writings, Volume 3: 1935–1938.* Edited by Howard Eiland and Michael Jennings. Cambridge, MA: Belknap Press of Harvard University Press.

Brueggemann, Walter. 2018. *The Prophetic Imagination: 40th Anniversary Edition.* Minneapolis, MN: Fortress Press.

Brueggemann, Walter. 2019. *From Judgment to Hope: A Study on the Prophets.* Louisville, KY: Westminster John Knox Press.

Buell, Lawrence. 1995. *The Environmental Imagination: Thoreau, Nature Writing, and the Formation of American Culture.* Cambridge, MA: Harvard University Press.

Cumpston, Zena, Michael-Shawn Fletcher, and Lesley Head. 2022. *Plants: Past, Present and Future.* Port Melbourne: Thames & Hudson Australia.

Deleuze, Gilles, and Felix Guattari. 1986. *Kafka: Towards a Minor Literature.* Translated by Dana Polan. Minneapolis: University of Minnesota Press. Frist published in French in 1975.

Gammage, Bill, and Bruce Pascoe. 2021. *Country: Future Fire, Future Farming.* Port Melbourne: Thames & Hudson Australia.

Giblett, Rod. 1996. *Postmodern Wetlands: Culture, History, Ecology.* Edinburgh: Edinburgh University Press.

Giblett, Rod. 2007. Black and White Water. In *Fresh Water: New Perspectives on Water in Australia*, ed. Emily Potter, Alison Mackinnon, Stephen McKenzie, and Jennifer McKay, 31–43. Carlton: Melbourne University Press.

Giblett, Rod. 2009. *Landscapes of Culture and Nature.* Basingstoke: Palgrave Macmillan.

Giblett, Rod. 2011. *People and Places of Nature and Culture.* Bristol: Intellect Books.

Giblett, Rod. 2013. *Black Swan Lake: Life of a Wetland.* Bristol: Intellect Books.

Giblett, Rod. 2016. *Cities and Wetlands: The Return of the Repressed in Nature and Culture.* London: Bloomsbury.

Giblett, Rod. 2019. *Environmental Humanities and the Uncanny: Ecoculture, Literature and Religion*. London: Routledge.

Giblett, Rod. 2020a. Artist and Swamp: Wetlands in Australian Painting and Photography. In *Australian Wetland Cultures: Swamps and the Environmental Crisis*, ed. John Charles Ryan and Li Chen, 51–70. Lanham, MD: Lexington Books.

Giblett, Rod. 2020b. *Modern Melbourne: City and Site of Nature and Culture*. Bristol: Intellect Books.

Giblett, Rod. 2021a. *Black Swan Song: Life and Work of a Wetland Writer*. Lanham, MD: Hamilton Books.

Giblett, Rod. 2021b. *Wetlands and Western Cultures: Denigration to Conservation*. Lanham, MD: Lexington Books.

Giblett, Rod. 2024. *Wetland Cultures: Ancient, Traditional, Contemporary*. New York: Palgrave Macmillan.

Hayes, E. R. 2012. Justice, Righteousness. In *Dictionary of the Old Testament: Prophets*, ed. Mark J. Boda and J. Gordon McConville, 466–472. Downers Grove, IL: InterVarsity Press.

Heschel, Abraham J. 2001. *The Prophets*. New York: Harper Perennial Modern Classics. First published in 1962.

Jefferies, Richard. 2001. In *At Home on the Earth: A New Selection of the Later Writings*, ed. Jeremy Hooker. Totnes: Green Books.

Jefferies, Richard. 2002. *The Story of my Heart: My Autobiography*. Totnes, Devon: Green Books. First published 1883.

Langton, Marcia. 1998. *Burning Questions: Emerging Environmental Issues for Indigenous Peoples in Northern Australia*. Darwin: Centre for Indigenous Natural and Cultural Resource Management, Northern Territory University.

Langton, Marcia. 2000. Homeland: Sacred Visions and the Settler State. *Artlink: Australian Contemporary Art Quarterly* 20 (1): 11–16.

Langton, Marcia. 2021. *Welcome to Country: A Travel Guide to Indigenous Australia*. 2nd ed. Richmond: Hardie Grant Explore.

Murdoch, Lindsay. 2010. *Kakadu being poisoned by Rio Tinto mine, group warns*. Sydney Morning Herald, May 24. Accessed online 3rd June 2023: https://www.smh.com.au/national/kakadu-being-poisoned-by-rio-tinto-mine-group-warns-20100523-w42y.html

Neidjie, Bill. 1989. *Story about Feeling*. Broome: Magabala Books.

Pascoe, Bruce, and Vicky Shukuroglou. 2020. *Loving Country: A Guide to Sacred Australia*. Richmond: Hardie Grant Travel.

Rose, Deborah Bird. 1988. Exploring an Aboriginal Land Ethic. *Meanjin* 47 (3): 378–387.

Scholem, Gershom. 2012. *On Jews and Judaism in Crisis: Selected Essays*. Translated by Werner J. Dannhauser. Philadelphia, PA: Paul Dry.

Thoreau, Henry David. 1958. *The Correspondence*. Edited by Walter Harding and Carl Bode. New York: New York University Press.

Thoreau, Henry David. 1962. *The Journal 1837–1861, Volumes I–XIV*. Edited by Bradford Torrey and Francis Allen. New York: Dover.

Thoreau, Henry David. 1982. *The Portable Thoreau*. Edited by Carl Bode. New York: Viking Penguin.

Thoreau, Henry David. 1998. *A Week on the Concord and Merrimack Rivers*. Edited by H. Daniel Peck. New York: Penguin.

Thoreau, Henry David. 2004. *Letters to a Spiritual Seeker*. Edited by Bradley Dean. New York: Norton.

Wood, David F. 2006. *An Observant Eye: The Thoreau Collection at the Concord Museum*. Concord, MA: Concord Museum.

CHAPTER 4

Nature Writing Is Prosaic and Poetic

Nature writing is written in prose. The anthologies and studies of nature writing exclude nature poetry, which would be an entirely different anthology and study. Some books and essays of nature writing written largely in prose include some poetry. Such is the case with Thoreau's *Walden* and his essays that include some poetry, his own, and others (as we will see in the present and later chapters). His prose and poetry include poetic figures of speech, such as metaphors and similes. He wrote poetry, as well as poetic and prosaic prose. Nature writing is about nature; it is also about the ordinary stuff of everyday life. Nature writing is prosaic in this sense of being about the ordinary and everyday life. Nature is ordinary. Gary Saul Morson and Caryl Emerson develop this double sense of prosaic as prose and ordinary in their study of the critical theory of Mikhail Bakhtin, the Russian cultural critic and literary theorist, early Soviet semiotician and internal exile under Stalin's regime (Morson and Emerson 1990, 10, 15). They follow in Bakhtin's footsteps using his term, the "prose of everyday life" (cited by Morson and Emerson 1990, 40).

Morson and Emerson also follow in Bakhtin's footsteps by privileging the genre of the novel as the prime instance of the prosaic in this sense and in the sense of being written in prose. They also argue against poetics and for prosaics for the appreciation and study of prose in its own right not relegated to secondary status after poetry (Morson and Emerson 1990, 15, 85, 90, 317–319). In this chapter I argue that nature writing is an

R. Giblett, *The Politics of Nature and Environmental Writing*, Transdisciplinary Environmental Humanities 1,
https://doi.org/10.1007/978-3-032-04427-3_4

equally estimable and valuable instance of the prosaic in both senses. I also argue for a prosaics and poetics of nature writing that appreciates the poetry and prose of nature writing, both poetic prose and prose poetry. What Bakhtin, Morson, and Emerson have to say about the novel also applies a great deal to nature writing (as we will see in this chapter and later ones). "The genre of the novel" for Bakhtin, Morson, and Emerson (1990, 17) is "a style of styles, an orchestration of the diverse language of everyday life into a heterogeneous sort of whole." The same could be said of the genre of nature writing, especially Thoreau's *Walden*, that points to its multi-generic characteristics and heterogenetic provenance like the novel (discussed in Chap. 6).

The prosaic also points to the messiness of everyday life. For Morson and Emerson (1990, 30) "the natural state of things is *mess*" (their emphasis). For them and Bakhtin mess is "the normal and at times even the healthy state" of culture (Morson and Emerson 1990, 30). Prosaic nature writing also points to the messiness of nature and culture, and the prosaic nature of everyday life. Nature writing is creative and creative writing. It is not usually taught in creative writing courses though, unless it might come under the purview of 'creative non-fiction' and be regarded as a sub-branch of it. For Morson and Emerson (1990, 187) "creativity is built into prosaic experience. [...] To live is to create." Creativity for them is "real, ongoing, and immanent in the process of daily living" (Morson and Emerson 1990, 298).

4.1 The Poetry and Prose of Poetic and Prosaic House-Building

Take, for instance, Thoreau's recounts of building and cleaning his house by Walden Pond interspersed throughout *Walden*. If Thoreau had given a straightforward, chronological account of building his house instead of digressing to discuss nature and sundry other topics, *Walden* may have sold better on its first publication as an owner-builder construction guide, but it would have been long out of print now and it would not have become the classic of nature writing that it has become, still be in print in many editions, and been translated into many languages. Thoreau weaves together the prose and prosaic activity of house-building by Walden Pond in Walden Woods and living in his house with the poetry of song and poetics of nature, especially the sounds of bird, with his own poetry and poetics of metaphor and simile. Ordinary, everyday life is elevated to the level of the extraordinary life of nature, and extraordinary nature is associated with ordinary, everyday life. Nature is ordinary and extraordinary. Nature is part and parcel of ordinary, everyday life. This is a typical and defining feature of Thoreau's writing (as we will see in this chapter).

Thoreau begins *Walden* in a prosaic way:

> When I wrote the following pages, or rather the bulk of them, I lived alone, in the woods, a mile from any neighbor, in a house which I had built myself, on the shore of Walden Pond, in Concord, Massachusetts, and earned my living by the labor of my hands only. I lived there two years and two months. At present I am a sojourner in civilized life again. (Thoreau 1982, 258)

It is not entirely true that he built the house by himself as he had help with raising the frame, though he also wrote that he didn't really need it. Thoreau blurred the line between fact and fiction, here as elsewhere.

After digressing from house-building for thirty-five pages, he returns to the prosaic topic of building his house:

> Near the end of March, 1845, I borrowed an axe and went down to the woods by Walden Pond, nearest to where I intended to build my house, and began to cut down some tall, arrowy white pines, still in their youth, for timber. (Thoreau 1982, 295)

Borrowing an axe prompts a meditation on the topics of borrowing and beginning ("it is difficult to begin without borrowing"), which applies equally to building a house as to writing a book (as I have attested elsewhere and as Edward Said studied; Giblett 2025, 188 n.1; Said 1978). Thoreau did both. Both are prosaic creative processes. When the axe head fell off, he inserted a new wedge into the handle, soaked it in water for it to expand, and returned the axe to its owner in better condition than that in which he had borrowed it. Returning a borrowing in better condition than which it was lent applies to writing a book as well as building a house (Giblett 2025, 188 n.1). The prose of everyday life becomes the philosophy of everyday life. This is a hallmark of Thoreau's nature writing.

Thoreau's account of house-building is interspersed with accounts of nature and with poetry, and sometimes immediately goes on to return to house-building:

> On the 1st of April it rained and melted the ice, and in the early part of the day, which was very foggy, I heard a stray goose groping about over the pond and cackling as if lost, or like the spirit of the fog.
>
> So I went on for some days cutting and hewing timber, and also studs and rafters, all with my narrow axe, not having many communicable or scholar-like thoughts, singing to myself,—
>
> Men say they know many things;
> But lo! they have taken wings—
> The arts and sciences,
> And a thousand appliances;
> The wind that blows
> Is all that any body knows.
>
> I hewed the main timbers six inches square, most of the studs on two sides only, and the rafters and floor timbers on one side, leaving the rest of the bark on, so that they were just as straight and much stronger than sawed ones. Each stick was carefully mortised or tenoned by its stump, for I had borrowed other tools by this time. (Thoreau 1982, 296–297)

Thoreau's nature writing quotes his own poetry. The prose and philosophy of everyday life includes house-building and being attentive to a stray goose displaced in the orderly course of nature. The prose of house-building in everyday life is also becoming the poetics of nature with the poetry of song and the simile of the cackling goose "like the spirit of the fog." Ordinary everyday life is elevated to the extraordinary life of nature, and extraordinary nature associated with the ordinary. Nature is part and parcel of ordinary, everyday life. This is a typical and defining feature of Thoreau's writing.

Thoreau then proceeds to an intermittent month-by-month account of the prosaic progress of his house-building:

> By the middle of April, for I made no haste in my work, but rather made the most of it, my house was framed and ready for the raising. I had already bought the shanty of James Collins, an Irishman who worked on the Fitchburg Railroad, for boards [...]. The roof was the soundest part, though a good deal warped and made brittle by the sun. (Thoreau 1982, 297)

April ends and so, "at length, in the beginning of May, with the help of some of my acquaintances, rather to improve so good an occasion for neighborliness than from any necessity, I set up the frame of my house" (Thoreau 1982, 299). The ever self-reliant Thoreau may have preferred to have built his house by himself to boast of it as he did in *Walden* (as we have seen and as he did again later; Thoreau 1982, 258, 303). June seems to have been largely spent in planting and hoeing beans. He later harvested and derived an income from them as a cash-crop and so indeed earnt his living by the labour of his hands (Thoreau 1982, 258, 412).

After three months of work and without going into any more detail about the process of house-building Thoreau proclaims:

> I began to occupy my house on the 4th of July, as soon as it was boarded and roofed, for the boards were carefully feather-edged and lapped, so that it was perfectly impervious to rain, but before boarding I laid the foundation of a chimney at one end, bringing two cartloads of stones up the hill from the pond in my arms. (Thoreau 1982, 299–300)

That's a lot of trips to the pond to carry that amount of stones. Taking stones from the pond would be frowned upon nowadays and be illegal in Walden Pond State Reservation in Massachusetts. Pilgrims to the site of Thoreau's house find and place a stone on the cairn located there. Thoreau seems to have resisted the temptation to moralize on the need to lay a good foundation in life as in house-building. He later advocated the virtues of finding and having "a hard bottom" (as we will see in Chap. 7).

The chimney was not completed when he occupied his house. It took several months for it to be completed:

> Before winter I built a chimney, and shingled the sides of my house, which were already impervious to rain, with imperfect and sappy shingles made of the first slice of the log, whose edges I was obliged to straighten with a plane. I have thus a tight shingled and plastered house, ten feet wide by fifteen long, and eight-feet posts, with a garret and a closet, a large window on each side, two trap doors, one door at the end, and a brick fireplace opposite. (Thoreau 1982, 303)

The precise and prosaic details of the measurements and components of his house were important for constructing the replica located at the Walden Pond Visitors Center and for the Unabomber reconstructing it for himself in Montana.

The chimney itself was built out of upcycled bricks, as he relates much later in *Walden*:

> When I came to build my chimney I studied masonry. My bricks, being second-hand ones, required to be cleaned with a trowel, so that I learned more than usual of the qualities of bricks and trowels. The mortar on them was fifty years old, and was said to be still growing harder; but this is one of those sayings which men love to repeat whether they are true or not. Such sayings themselves grow harder and adhere more firmly with age, and it would take many blows with a trowel to clean an old wiseacre of[f] them. (Thoreau 1982, 485)

The point of this passage is not about chimney construction and its material, nor about Thoreau's ongoing recount of house-building as he had completed it, including the chimney, and was living in it. It is about old clichés and re-assuring sayings that have a phatic function at which both parties of teller and listener nod sagely. Whether or not the mortar had hardened was beside the point for Thoreau who leverages the prose of his everyday life of house-building at the time into a poetic discussion of 'wiseacres' in the lives of others, irrespective of whether they are true or not too. Prosaic objects of everyday life—bricks and mortar—are the pre-text for critique about the role of 'wiseacres' in the lives of others where they are often repeated whether they are true or not. Thoreau preferred true wiseacres, the ones he tested and tried in everyday life. He also preferred writing his own wise sayings that have become "quotable quotes," plenty of which have been collected into a single volume of nearly 500 pages (Cramer 2011).

After relating his earlier chimney-building and digressing from house-building for another thirty-five pages, he returns to this prosaic topic and repeats the momentous date of his occupancy:

> When first I took up my abode in the woods, that is, began to spend my nights as well as days there, which, by accident, was on Independence Day, or the Fourth of July, 1845, my house was not finished for winter, but was merely a defence against the rain, without plastering or chimney, the walls being of rough, weather-stained boards, with wide chinks, which made it cool at night. (Thoreau 1982, 337–338)

History and geography intersect in this place. Thoreau's house, and his building of it, is his declaration of independence, after being dependent on Ralph Waldo Emerson for the location and the timber, on the Collins for the recycled materials, and acquaintances (not friends) for the borrowed tools and labouring assistance in raising the frame. Thoreau (1982, 486) later plastered his house and made it snug (Thoreau 1982, 418). Writing a book and revising it for publication is filling the chinks in rough drafts with a word or two here or there, or a sentence too, to stop the winds of woe blowing through it, and make it a defence against the draughts of despair and the slings and arrows of outrageous fortune.

Thoreau took up residence as the lord of his modest manor and master of all that he surveyed on that momentous day commemorating the, and his, declaration of independence. The prose of everyday life becomes the poetics of place, its geography, and history in his psychopolitical ecology:

> I was seated by the shore of a small pond, about a mile and a half south of the village of Concord and somewhat higher than it, in the midst of an extensive wood between that town and Lincoln, and about two miles south of that our only field known to fame, Concord Battle Ground; but I was so low in the woods that the opposite shore, half a mile off, like the rest, covered with wood, was my most distant horizon. (Thoreau 1982, 339)

Thoreau's modest one-room house built in and from Emerson's wood lot invites Thoreau to play on 'seat' and parody a country gentleman's 'seat' or estate. The locality and its history have a claim to national fame as the first place where a shot was fired in the American Revolutionary War commemorated on Patriots Day in April every year in Massachusetts and five other states.

The prose of everyday life of house-building and a declaration of independence and of revolutionary war also becomes the prose of personal development and education, a non-fiction *Bildungsroman*, the novel of emergence:

> I went to the woods because I wished to live deliberately, to front only the essential facts of life, and see if I could not learn what it had to teach, and not, when I came to die, discover that I had not lived. I did not wish to live what was not life, living is so dear; nor did I wish to practise resignation, unless it was quite necessary. I wanted to live deep and suck out all the marrow of life, to live so sturdily and Spartan-like as to put to rout all that was not life, to cut a broad swath and shave close, to drive life into a corner, and reduce it to its lowest terms, and, if it proved to be mean, why then to get the whole and genuine meanness of it, and publish its meanness to the world; or if it were sublime, to know it by experience, and be able to give a true account of it in my next excursion. (Thoreau 1982, 343)

For the nature writer; living meant going on excursions and giving an account of them (as Thoreau did on many occasions). Thoreau wanted to sublimate his prosaic everyday life into either the heroic poetics of a Spartan life or the sublime philosophy of the secular sacred. Or both in his psychopolitical ecology. Thoreau sought to live both the active life of going on excursions and the contemplative life of theology and philosophy, and to recount the combination of them in his poetic prose.

4.2 The Poetry and Prose of Poetic and Prosaic Housework

Living meant staying at home, performing the mundane chores of everyday life, and recounting the prosaic in poetic prose:

> Housework was a pleasant pastime. When my floor was dirty, I rose early, and, setting all my furniture out of doors on the grass, bed and bedstead making but one budget, dashed water on the floor, and sprinkled white sand from the pond on it, and then with a broom scrubbed it clean and white. (Thoreau 1982, 364–365)

Prosaic house-cleaning prompts Thoreau to boast that he got his housework done before "the villagers" of Concord had their breakfast. He could not resist having a sly dig at them and highlighting his point of difference from them. After returning his furniture to his house, he commented that "my meditations were almost uninterrupted" (Thoreau 1982, 365). Thoreau did not perform his meditations uninterrupted while he was doing his housework.

Thoreau's meditations were performed at other times, in other activities, preferably seated on his threshold:

> Sometimes, in a summer morning, having taken my accustomed bath, I sat in my sunny doorway from sunrise till noon, rapt in a revery, amidst the pines and hickories and sumachs, in undisturbed solitude and stillness, while the birds sing around or flitted noiseless through the house, until by the sun falling in at my west window, or the noise of some traveller's wagon on the distant highway, I was reminded of the lapse of time. I grew in those seasons like corn in the night, and they were far better than any work of the hands would have been. They were not time subtracted from my life, but so much over and above my usual allowance. (Thoreau 1982, 363–364)

Thoreau practised silent sitting, cultivated stillness, and dreamt awake in a reverie in his psychopolitical ecology. He meditated in human solitude in the company of birds accompanied by their song. He lost track of time rapt in his reverie (what he also called "dreaming awake," as we will see in a later chapter). He grew like a seed in the dark in the subterranean soil or cellar.

Thoreau could while away not only a morning but also an afternoon and an evening:

> Some of my pleasantest hours were during the long rain-storms in the spring or fall, which confined me to the house for the afternoon as well as the forenoon, soothed by their ceaseless roar and pelting; when an early twilight ushered in a long evening in which many thoughts had time to take root and unfold themselves. (Thoreau 1982, 383)

The seed germinating in the dark had time to take root, germinate, grow, and unfold like a plant. The prosaics of everyday life is married to the poetics of nature in psychopolitical ecology.

Such was also the case with the sounds of birds:

> For sounds in winter nights, and often in winter days, I heard the forlorn but melodious note of a hooting owl indefinitely far; such a sound as the frozen earth would yield if struck with a suitable plectrum, the very *lingua vernacula* of Walden Wood. (Thoreau 1982, 514)

Thoreau has ears to hear the ordinary utterances of the beings who people the Wood that are sung in their music and his poetics of metaphor. Describing or reproducing the sounds of birds is a prominent feature of Thoreau's nature writing. These are repeatedly mentioned in a recent anthology of his writing about birds collected from his journal (Wisener 2024). He had an attentive ear for "listening *intentionally*" to hear the sounds of birds, as he wrote in his journal (Thoreau 1962, *XII,* 42; his emphasis; cited by Wisener 2024, 6). He recorded and wrote about them in a rich vein of poetic language, especially onomatopoeia, as nearly every page of the recent anthology shows.

From discussing the hooting owl, Thoreau goes on in *Walden* to relate that he hears the honking of goose leading a flock of geese, the sound of whose wings is "like a tempest in the woods," only to be interrupted by an owl:

> One night in the beginning of winter, before the pond froze over, about nine o'clock, I was startled by the loud honking of a goose, and, stepping to the door, heard the sound of their wings like a tempest in the woods as they flew low over my house. They passed over the pond toward Fair Haven, seemingly deterred from settling by my light, their commodore honking all the while with a regular beat. Suddenly an unmistakable cat-owl from very near me, with the most harsh and tremendous voice I ever heard from any inhabitant of the woods, responded at regular intervals to the goose, as if determined to expose and disgrace this intruder from Hudson's Bay by exhibiting a greater compass and volume of voice in a native, and *boo-hoo* him out of Concord horizon. What do you mean by alarming the citadel at this time of night consecrated to me? Do you think I am ever caught napping at such an hour, and that I have not got lungs and a larynx as well as yourself? *Boo-hoo, boo-hoo, boo-hoo!* It was one of the most thrilling discords I ever heard. And yet, if you had a discriminating ear, there were in it the elements of a concord such as these plains never saw nor heard. (Thoreau 1982, 514–515)

Even the discord of the harsh and tremendous voice of the owl is a singer in the choir of Concord, just as much as the honking of a goose and the tempest of geese wings for those attentive ears who listen intentionally and hear the sounds of nature in the prose and poetry of ordinary and extraordinary life in psychopolitical ecology.

4.3 The Poetry and Prose of Poetic and Prosaic Walking and Writing

Prose and poetry, everyday life and nature, come together in walking and writing. Walking is an elementary practice of everyday life, as Michel de Certeau (1984, 93) argued. Walking is prosaic. Thoreau is famous as a walker, a chronicler of his own walking, and a philosopher of walking in the country. Rambling, botanizing, and zoologizing nature in the countryside were popular nineteenth-century pastimes. Thoreau refers to "my botanical rambles" in his journal and late essay "Autumnal Tints" (Thoreau 1962, *XI*, 285; 2001, 393). He uses variations of "ramble" and "the rambler" eight times in *Walden*, including a critique of his contemporaries when they use these terms idly, fashionably, and superficially. Observing rocks, plants, and animals, and occasionally collecting specimens of them, were everyday activities for Thoreau. He developed a walking ecology of the country. He was what Buell (1995, 389) calls "nature's *flâneur*," though culture was everywhere present in the traces of first nations and the intrusions of the modern industrial capitalist technology of the railroad on Walden Pond and in Walden Woods. He was nature's and culture's *flâneur.*

Rambling also has a pejorative connotation of a peripatetic and roundabout style of writing. It is an apt descriptor sometimes for Thoreau's style of walking and writing, though on other occasions they are also direct and stride forth in one direction in rhythm with each other. Writing and walking go hand-in-hand, foot-by-foot. As Thoreau walked, so he wrote and vice versa. Walking and thinking for Thoreau also went in step and in time with each other. Thoreau (1962, *II,* 404–405) wrote in his journal "that the moment my legs begin to move, my thoughts begin flow." He wrote much later in his journal that "a man thinks as well through his legs and through his brain. We exaggerate the importance and exclusiveness of the headquarters" (Thoreau, 1962, *XIII,* 69–70). Thoreau emphasized the importance and inclusiveness of the hind-quarters too.

Thoreau preferred sauntering to rambling. Sauntering freely is good for physical and mental or spiritual health. Thoreau prescribes and practices the walking cure for environ-mental illness and for environ-mental health. He also engages in the talking cure by playing freely with words and their associations when he concocts a dubious, but poetic, etymology for the word "saunter" from "s*ainte-Terrer,*" a "Holy Lander," a pilgrim to the Holy Land. Sauntering for Thoreau is a holy act on holy land. Walking and its eponymous essay are a spiritual journey. Thoreau also links saunter to "s*ans terre,*" "without land or home [...] having no particular home, but equally at home everywhere" (Thoreau 1982, 592–593; 2001, 225).

Thoreau still loved to return home to his house after he had left it to go sauntering:

> It was very pleasant, when I stayed late in town, to launch myself into the night, especially if it was dark and tempestuous, and set sail from some bright village parlor or lecture room, with a bag of rye or Indian meal upon my shoulder, for my snug harbor in the woods, having made all tight without and withdrawn under hatches with a merry crew of thoughts, leaving only my outer man at the helm, or even tying up the helm when it was plain sailing. (Thoreau 1982, 418)

"Now comes good sailing" are reputedly Thoreau's four last words whispered to his sister on his deathbed. They are also the title of a recent anthology of writers reflecting on Thoreau (Blauner 2022). Death was plain sailing for Thoreau into a snug harbour, just as Thoreau the walker in the woods at night sailed the dark seas to his house harbour.

Thoreau did not get lost on such voyages walking home to his house harbour by the pond as his eyes and feet knew the way:

> It is remarkable how easily and insensibly we fall into a particular route, and make a beaten track for ourselves. I had not lived there a week before my feet wore a path from my door to the pond-side; and though it is five or six years since I trod it, it is still quite distinct. It is true, I fear, that others may have fallen into it, and so helped to keep it open. The surface of the earth is soft and impressible by the feet of men; and so with the paths which the mind travels. How worn and dusty, then, must be the highways of the world, how deep the ruts of tradition and conformity! I did not wish to take a cabin passage, but rather to go before the mast and on the deck of the world, for there I could best see the moonlight amid the mountains. I do not wish to go below now. (Thoreau 1982, 562)

Thoreau spent two years before the mast in the woods, like Richard Henry Dana who spent two years before the mast at sea and recounted it in *Two Years Before the Mast* published in 1840.

By night the woods enclosed him on his walk home:

> It is darker in the woods, even in common nights, than most suppose. I frequently had to look up at the opening between the trees above the path in order to learn my route, and, where there was no cart-path, to feel with my feet the faint track which I had worn, or steer by the known relation of particular trees which I felt with my hands, passing between two pines for instance, not more than eighteen inches apart, in the midst of the woods, invariably, in the darkest night. Sometimes, after coming home thus late in a dark and muggy night, when my feet felt the path which my eyes could not see, dreaming and absent-minded all the way, until I was aroused by having to raise my hand to lift the latch, I have not been able to recall a single step of my walk, and I have thought that perhaps my body would find its way home if its master should forsake it, as the hand finds its way to the mouth without assistance. (Thoreau 1982, 418–419)

He was dreaming awake while walking and did not get lost because his body, or his autonomic nervous system as we would say these days, remembered the way home. He goes on to relate several incidents of those walkers who got lost at night in the woods, and even in the village when "the darkness was so thick that you could cut it with a knife, as the saying is" (Thoreau 1982, 419). Thoreau did not get lost in the woods at night because they were home and the way to his snug parlour.

Thoreau at home was enclosed in his snug parlour:

> My dwelling was small, and I could hardly entertain an echo in it; but it seemed larger for being a single apartment and remote from neighbors. All the attractions of a house were concentrated in one room; it was kitchen, chamber, parlor, and keeping-room; and whatever satisfaction parent or child, master or servant, derive from living in a house, I enjoyed it all. (Thoreau 1982, 487)

Perhaps the present-day residents of "studio" apartments might not be so effusive about having their kitchen, bedroom, living room, and pantry all rolled into one room. Henri Pollès (cited by Benjamin 1999, 225) observed in 1837 that the marketing of such apartments as studios was "as if people were more and more becoming artists or students," or as if more and more people were becoming artists or students. In Thoreau's case, he was already an artisan and a student. He described his one-room house as "at once kitchen, pantry, parlor, chamber, storehouse, and garret" (Thoreau 1982, 488). It was a kitchen for preparing and eating food, a pantry for storing food and utensils, a parlour for receiving and entertaining visitors, a bedroom for sleeping and dreaming asleep or awake, a garret for a poet to write in, and a storehouse for his library of journals and a couple of books. Thoreau (1982, 489) bemoans the fact that "the parlor is so far from the kitchen and workshop" in some houses of his day with slides and dumbwaiters compared to his house in which the parlour, kitchen, and workshop are all rolled into one room. The Woods and the Pond are not only the location of his house, but also his home, his home place, and homes also for others as the pond is "the parlor of the fishes" (Thoreau 1982, 525).

The walker walking away from home and returning back home in a loop is a circumnavigator of their own walking world, a Ferdinand Magellan of the land. Thoreau counsels setting forth on journeys of exploration into the great unknown within oneself when he asks "[i]s not our own interior white on the chart? black though it may prove, like the coast, when discovered. […B]e a Columbus to whole new continents and worlds within you, opening new channels, not of trade, but of thought. […O]bey the precept of the old philosopher, and Explore thyself" (Thoreau 1982,

560–561). The Greek philosophical maxim is actually "know thyself." Walter Harding notes in his annotated edition of *Walden* that Thoreau was alluding to this apothegm (Thoreau 1995, 313 n6). Yet "explore" and "know" are not synonymous. Thoreau rejigs and repurposes the maxim to give priority to exploration over knowledge, *εξερεύνηση* over *η γνώση*. One cannot know oneself if one has not already explored oneself, especially those regions of oneself that are white blanks on the chart and need filling in, even though they may prove to be the heart of darkness in the dark continent of the unconscious, the grotesque lower mental stratum (discussed in Chap. 7). One cannot know oneself or anything else for Thoreau until one has forgotten one's learning. He wrote in his journal that "it is only when we forget all our learning that we begin to know" (Thoreau 1962, *XII*, 371). We only begin to know ourselves if and when we have explored ourselves. The prosaic and poetic nature writing of Thoreau is a guide for doing so.

References

Benjamin, Walter. 1999. *The Arcades Project*. Translated by Howard Eiland. Cambridge, MA: Belknap Press of Harvard University Press.

Blauner, Andrew, ed. 2022. *Now Comes Good Sailing: Writers Reflect on Henry David Thoreau*. Princeton, NJ: Princeton University Press.

Buell, Lawrence. 1995. *The Environmental Imagination: Thoreau, Nature Writing, and the Formation of American Culture*. Cambridge, MA: Harvard University Press.

Certeau, Michel de. 1984. *The Practice of Everyday Life*. Translated by Steven Rendall. Berkeley, CA: University of California Press.

Cramer, Jeffrey, ed. 2011. *The Quotable Thoreau*. Princeton, NJ: Princeton University Press.

Giblett, Rod. 2025. *Waking to Nature with Thoreau and Benjamin: Or, Psychopolitical Ecology*. New York: Bloomsbury.

Morson, Gary Saul, and Caryl Emerson. 1990. *Mikhail Bakhtin: Creation of a Prosaics*. Stanford: Stanford University Press.

Said, Edward. 1978. *Beginnings: Intentions and Method*. Baltimore, MD: Johns Hopkins University Press.

Thoreau, Henry David. 1962. *The Journal 1837–1861, Volumes I–XIV*. Edited by Bradford Torrey and Francis Allen. New York: Dover.

Thoreau, Henry David. 1982. *The Portable Thoreau*. Edited by Carl Bode. New York: Viking Penguin.

Thoreau, Henry David. 1995. *Walden: An Annotated Edition, Foreword and notes by Walter Harding*. Boston: Houghton Mifflin.

Thoreau, Henry David. 2001. *Collected Essays and Poems*. New York: Library of America.

Wisener, Geoff, ed. 2024. *A Year of Birds: Writings on Birds from the Journal of Henry David Thoreau*. Macon, GA: Mercer University Press.

CHAPTER 5

Nature Writing Is Dialogic and Polyphonic

Nature writing engages in dialogue between the human writer and more-than-human plants, and animals, and their places and processes. Mikhail Bakhtin is the leading critical theorist of the dialogic mainly in his study of Dostoevsky (Bakhtin 1984) and signalled in the title of the collection of his essays, *The Dialogic Imagination* (Bakhtin 1981), yet less so in the four essays collected in it. The dialogic imagination is celebrated and performed in what he called elsewhere the creative, carnivalesque, and intersubjective word and in the grotesque body. It is distinct from the singular monologic and monotheistic conceptualization dictated in destructive, disembodied, and objective discourses (institutionalized practices producing significance) pronounced by *the* subject. The exemplar is the fascist, totalitarian, or authoritarian single-voice speech of an autocratic and dictatorial president, commander-in-chief, or supreme ruler.

5.1 Nature Writing Is Dialogic

For Bakhtin "life by its very nature is dialogic" (cited by Morson and Emerson 1990, 60). Bakhtin (1984, 252) argued in his study of Dostoevsky that "to be means to communicate dialogically. When dialogue ends, everything ends." Bakhtin for Morson and Emerson (1990, 59) in their study of his critical theory "envisaged all of life as an ongoing, unfinalizable dialogue." Nature writing is dialogic as it engages in dialogue between

R. Giblett, *The Politics of Nature and Environmental Writing*, Transdisciplinary Environmental Humanities 1, https://doi.org/10.1007/978-3-032-04427-3_5

humans and more-than-humans. It also celebrates the dialogue between nature and humans. What Bakhtin (1984, 252) wrote of Dostoevsky applies to Thoreau that he "carries dialogue into eternity, conceiving of it as eternal co-rejoicing, co-admiration, con-cord." Thoreau lived in and near the town of Concord most of his life and wrote dialogically in concord with nature, and "in the nick of time" between the eternity of the past and the future.

Thoreau wrote in his journal that "I find it good to remember the eternity behind me as well as the eternity before" (Thoreau 1962, *I,* 337). He wrote in *Walden* that the eternal past and future meet in the present moment when "in any weather, at any hour of the day or night, I have been anxious to improve the nick of time, and notch it on my stick too; to stand on the meeting of two eternities, the past and future, which is precisely the present moment; to toe that line" (Thoreau 1982, 272). Thoreau was using a combined spatial and temporal figure as he literally nicked his walking stick with the marks of inches and used it as a portable ruler to take measurements (Wood 2006, 63–64). Rather than toeing the line for the boss or the state, Thoreau preferred toeing the line of the present moment as it is divine. For Thoreau later in *Walden* "God himself culminates in the present moment" (Thoreau 1982, 349). "God" is Thoreau's name for the culminator of the present moment. God is not a being outside time who created time (and everything else), but the culminator of the present moment of time itself.

What Bakhtin (1984, 252) also wrote of Dostoevsky and dialogic relationships applies to Thoreau. Dialogic relationships for Bakhtin

> are an almost universal phenomenon, permeating all human speech and all relationships and manifestations of human life-in general, everything that has meaning and significance. Dostoevsky could hear dialogic relationships everywhere, in all manifestations of conscious and intelligent human life; where consciousness began, there dialogue began for him as well. Only purely mechanistic relationships are not dialogic. (Bakhtin 1984, 40)

Thoreau could hear dialogic relationships everywhere too, in all manifestations of conscious and intelligent human and non-human life, such as the laughter of a loon on Walden Pond. The mechanistic relationships of the Fitchburg Railroad that "touches the pond" and whose "panting" interrupted Thoreau's dreams were not dialogic (Thoreau 1982, 367, 441, 632; 1962, *III,* 331).

Elsewhere for Bakhtin (1986, 111), and along the same lines, "there can be no dialogic relationship with an object." Dialogic relationships are between subjects; they are intersubjective. For him "in the process of dialogic communication, the object is transformed into the subject (the other's *I*)" (Bakhtin 1986, 145). Nature writing transforms the objects of nature (plants, animals, and places) into subjects and enters into intersubjective, occasionally abjective, relationship with them, whereas natural history always maintains monologic communication between the subject of the natural historian and the objects of nature. Bakhtin (1984, 68; his emphases) also suggested that "the consciousnesses of other people cannot be perceived, analyzed, defined as objects or as things—one can only *relate to them dialogically*." The consciousnesses of other beings, like loons, cannot be perceived, analysed, defined as objects or as things—one can only relate to them dialogically as Thoreau did.

Thoreau lived under the sign of what Walter Benjamin called "the new sort of researcher" who had "the ability to be at home in border regions" (cited by Jennings 1987, 155), as Benjamin himself lived. These border regions included the boundaries of genres, disciplines, times, places, and magic. Unlike Proust, Thoreau was more medieval than modern; like Benjamin, he was pre-secular and not post-secular. Benjamin argues in *The Arcades Project*:

It is a shock that brings someone engrossed in reverie up from the depths. Medieval legends invoke the state of shock peculiar to the researcher whose longing for more-than-human wisdom has led him to magic; the experience of shock is cited here as the "derisive laughter of hell": "Here … the muteness of matter is overcome. In laughter, above all, matter takes on an abundance of spirit, in highly eccentric disguise, embraces spirit. Indeed, it becomes so spiritual that it far outstrips [or greatly overshoots] language. Aiming still higher, it ends in shrill laughter." (Benjamin 1999a, 325)

Benjamin is quoting himself in his study of the German mourning play (Benjamin 2019, 248). Thoreau was led to the mute magic of nature where he is "a wizard," as he wrote in his journal (Thoreau 1962, *I*, 261). He also wrote later in his journal that "the physical fact […] in all language is the symbol of the spiritual" (Thoreau 1962, *XIII*, 145). Materiality and spirituality for Thoreau are unified in life and language.

A shock brings Thoreau engrossed in his reverie up from the depths floating on Walden Pond. Modern legends invoke the unique delight of Thoreau the researcher whose longing for more-than-human wisdom led him to the magic and wizardry of nature and to mocking laughter where the muteness of nature is overcome and matter takes on an abundance of spirit that outstrips language. Thoreau devotes many pages in *Walden* to chronicling a long story of a game of hide and seek that he played in his boat on the pond with a disappearing and re-appearing loon with the magical, mocking laughter of nature:

> In the fall the loon (*Colymbus glacialis*) came, as usual, to moult and bathe in the pond, making the woods ring with his wild laughter before I had risen. [...] [H]is unearthly laugh [...] was this demoniac laughter, yet somewhat like that of a water-fowl; but occasionally, when he had balked me most successfully and come up a long way off, he uttered a long-drawn unearthly howl, probably more like that of a wolf than any bird; as when a beast puts his muzzle to the ground and deliberately howls. This was his looning—perhaps the wildest sound that is ever heard here, making the woods ring far and wide. I concluded that he laughed in derision of my efforts, confident of his own resources. Though the sky was by this time overcast, the pond was so smooth that I could see where he broke the surface when I did not hear him. His white breast, the stillness of the air, and the smoothness of the water were all against him. At length having come up fifty rods off, he uttered one of those prolonged howls, as if calling on the god of loons to aid him, and immediately there came a wind from the east and rippled the surface, and filled the whole air with misty rain, and I was impressed as if it were the prayer of the loon answered, and his god was angry with me; and so I left him disappearing far away on the tumultuous surface. (Thoreau 1982, 479–482)[1]

Benjamin (1996, 303) argued that "to the attentive spectator nature is 'never dead or mute'," citing Goethe. Thoreau was certainly an attentive spectator with "an awakened eye," as Buell (2024, 48) puts it, ready to perceive, receive, and experience "nature's awakening," as Buell (2024, 46) also puts it. Thoreau was awake to alive nature and it awakened him in a dialogic, back and forth, to-and-fro process with nature. He was not only an attentive spectator and awakened eye but also an attentive listener

[1] Thoreau (1962, *IV*, 379–381) wrote about this game of hide and seek in his journal in 1852, five years after he had left his house by Walden Pond and while he was visiting the pond during the period he was writing, rewriting, and preparing *Walden* for publication in 1854. When Thoreau (1982, 337) claimed in *Walden* that he was "putting the experience of two years into one," the claim was in fact a fiction. He was the blurring the line between fiction and "non-fiction" here and elsewhere and crossing the great divide between them.

and awakened ear to whom nature spoke. Benjamin (1996, 73) argued earlier that "because she is mute, nature mourns." Arsić (2016, 379) suggests in a discussion of Thoreau, Freud, and mourning that Benjamin "followed a similar path to Thoreau" and that it is possible "in the whole (modern) history of Western thought, only Benjamin [...] came close to Thoreau's ideas of nature as endless/formless mourning that recreates as it grieves." It is also possible that only Arsić has seen the similarity between their views of language and "nature in mourning," as well as their similar "materialist epistemology" (Arsić 2016, 6, 252, 369), while overlooking their similar spiritual epistemology and engagement with theology. Arguably Thoreau's "vitalistic materialism" as distinct from "idealistic materialism" (Arsić 2016, 121–122) was also shared with Benjamin, as Benjamin's materialist theology was shared with Thoreau in psychopolitical ecology (Giblett 2025).

5.2 Nature Writing Is Polyphonic

Nature writing is also what Bakhtin called 'polyphonic,' or multi-voiced. The nature writer listens to and hears the voices of nature and conveys them to the reader, often quite literally with animals, especially birds. Nature writing also writes a word for nature that cannot speak itself, or only has a voice for those, such as nature writers, who can hear it and its plants, rocks, or waters speak. Dostoevsky for Bakhtin is 'the creator of the polyphonic novel' (Bakhtin 1984, 7). Thoreau is the creator of polyphonic nature writing.

Polyphony for Morson and Emerson (1990, 256) is "a theory of creativity"; it is also a practice of creativity. Bakhtin (1984, 201) argues that "for the prose artist the world is full of other people's words." For the nature writing prose artist the world is full of the other's words, not just people. Morson and Emerson (1990, 70) argue that Bakhtin defined "the novelistic *word* [...] as open, unfinalized, inhabited by many voices." The same could be said of Thoreau's nature writing word that includes him listening to and hearing the sounds of bird notes in song, the laugh of the loon, geese honking, ice cracking, a train snorting, etc. What Buell (1995, 397) calls Thoreau's "environmental nonfiction" was "in some ways more 'polyphonic,' more 'heteroglossic,' than that of the novel." Buell notes at this point that "Thoreauvian multigenericism" seems to him:

> to combine a greater array of genres and mimetic modes (e.g. 'fiction' and 'non-fictional' mimesis) and to create larger fissures within a text than the discursive universe of the novel as mapped by Bakhtin. (Buell 1995, 554 n.3)

"Thoreauvian multigenericism" and the polyphony of Thoreau's "environmental nonfiction" are as estimable and valuable as the multi-genericism and polyphony of the novel (as this book demonstrates). The prose artist is not only the novelist but also the nature writer who should be as highly regarded as the novelist.

Polyphony for Morson and Emerson (1990, 232) "has to do with the position of the author in a text." The author of nature writing often takes the position of the first-person narrator of the text. This is certainly the case with Thoreau when he writes in *Walden:*

> In most books, the *I*, or first person, is omitted; in this it will be retained; that, in respect to egotism is the main difference. We commonly do not remember that it is, after all, always the first person that is speaking. I should not talk so much about myself if there were anybody else whom I knew as well. Unfortunately, I am confined to this theme by the narrowness of my experience. Moreover, I, on my side, require of every writer, first or last, a simple and sincere account of his own life, and not merely what he has heard of other men's lives; some such account as he would send to his kindred from a distant land; for if he has lived sincerely, it must have been in a distant land to me. (Thoreau 1982, 259)

For Morson and Emerson (1990, 240) "the polyphonic world is Copernican." Dostoevsky for Bakhtin (1984, 49) "carried out, as it were, a small-scale Copernican revolution." So did Thoreau. The text and the world for the young Thoreau revolved around him. For the later Thoreau they no longer revolve around him as the centre, like the sun at the centre of the solar system. The centre of the writer's solo system has shifted from him to nature and the other beings in it of which he is one among many. Polyphony for Morson and Emerson (1990, 244) means that "the author's consciousness is but one of many consciousnesses." They conclude that "Dostoevsky's creation of polyphony constitutes a Copernican revolution" (Morson and Emerson 1990, 281).

What is important for Bakhtin (1984, 287; his emphases) with Dostoevsky is "not that which takes place within, but that which takes place on the *boundary* between one's own and someone else's consciousness, on the *threshold*." For the young Thoreau of *A Week on the Concord and Merrimack Rivers* what was important was what was within his own consciousness. For the older Thoreau of *Walden* what became important for him were other consciousnesses and the boundary between his and theirs, including the squirrels that came to eat on the threshold of his house and the swallow that landed on his shoulder.

The younger and older Thoreau can be compared, as Bakhtin compares Goethe and Dostoevsky:

> Goethe gravitates organically toward an evolving sequence. He strives to perceive all existing contradictions as various stages of some unified development; in every manifestation of the present he strives to glimpse a trace of the past, a peak of the present-day, or a tendency of the future. (Bakhtin 1984, 28)

The early Thoreau of *A Week* is like Goethe with the flow and confluence of the two rivers in which the traces of the life of his brother with whom he boated and was now dead when he wrote the book, the peaks of the present-day when he was writing the book, and the tendency of the future were organically unified.

Bakhtin goes on:

> In contrast to Goethe, Dostoevsky attempted to perceive the very stages themselves in their *simultaneity*, to *juxtapose* and *counterpose* them dramatically, and not stretch them out into an evolving sequence. For him, to get one's bearings on the world meant to conceive all its contents as simultaneous, and *to guess at their interrelationships in the cross-section of a single moment*. (Bakhtin 1984, 28; his emphases)

The later Thoreau of *Walden* and "Walking" is like Dostoevsky standing on the intersection of the past and the future in the present (as we have seen).

Along similar lines, time for Richard Jefferies is:

> eternity now. I am in the midst of it. It is about me in the sunshine; I am in it, as the butterfly floats in the light-laden air. Nothing has to come; it is now. Now is eternity; now is the immortal life. [...] To the soul there is not past and no future; all is and will be ever, in now. [...] The wonder is here, not there; now, not to be, now always. (Jefferies 2001, 39, 40)

Home is here, now, not there, then.

5.3 Thoreau's Water Bodies

Thoreau's Copernican revolution is evident in his views of water bodies that changed over time from his early river journey in the boys-own adventure story of *A Week*, through a Romanticist and Transcendentalist self-obsession with lakes and ponds in *Walden*, to a later sacral reverence for swamps in "Walking." Walter Benjamin (1996, 245, 303) in his essay on Goethe's *Elective Affinities* critiqued "the self-satisfied inwardness of Romanticism" manifested in "the lake" and "its unholy nature under the dead plane of the mirroring surface."[2] Self-satisfied inwardness is an annoying feature of some parts of Thoreau's writings, as many readers can attest.[3] For the early Thoreau the mirroring surface of lakes and ponds is a means for self-satisfied inwardness. It is also a means for mystical union with holy nature under the living plane of the mirroring surface of lakes. Thoreau also looked beyond himself into the heavens above and he looked below the surface of lakes into their and his depths. Thoreau (1962, *V*, 4) wrote in his journal that "I am a mystic, a transcendentalist, and a natural philosopher to boot." The lake and the pond for Thoreau were never unholy nature under the dead plane of the mirroring surface of lakes as it was for Romanticism.

He announced in *Walden* of Walden Pond that:

> this small lake was of most value as a neighbor in the intervals of a gentle rain-storm in August, when, both air and water being perfectly still, but the sky overcast, mid-afternoon had all the serenity of evening, and the wood thrush sang around, and was heard from shore to shore. A lake like this is never smoother than at such a time; and the clear portion of the air above it being shallow and darkened by clouds, the water, full of light and reflections, becomes a lower heaven itself so much the more important. (Thoreau 1982, 339–340)

The reflections in the water become a lower heaven under the higher heaven of the sky above. They bring heaven down to earth.

Thoreau also looked beyond and deeper below the reflections on the surface of the water into its and his depths when he later announced in

[2] In his essay "On the Image of Proust" Benjamin (1973, 215; 1999b, 245) repeated the phrase "the self-satisfied inwardness of Romanticism."

[3] Buell (2024) tackles the highs and lows, contradictions, cantankerousness, and crotchets of Thoreau in his recent meta-study.

Walden that "a lake is the landscape's most beautiful and expressive feature. It is earth's eye; looking into which the beholder measures the depth of his own nature" (Thoreau 1982, 435). Thoreau wrote in his journal and repeated it in his essay, "A Winter Walk," that "the lake is a mirror in the breast of nature. [...] It is the earth's liquid eye" (Thoreau 1958, 185; 1982, 67). He also described in his journal "the smooth reflecting surface of woodland lakes" as "those liquid eyes of his nature" (Thoreau 1962, *IV*, 147). He looks into the eye of the lake and surveys the depth of his own nature. He experiences what could be called the lacustrine ocular unconscious. Unlike "the optical unconscious" that Benjamin (1979, 243; 1999b, 510; 2003, 266) perceived in early photography and wrote about in several places in which "another nature speaks to the camera," Thoreau looking into the lake's eye experiences both the depth of his own nature and nature as other and akin, not alien.

When Thoreau (1982, 436) looks below the surface of Walden Pond that is not hard and solid (like a mirror) it is in a semi-liquid state it is "like molten glass cooled but not congealed, and the few motes in it are pure and beautiful like the imperfections in glass." Thoreau elaborated at length in *Walden*:

> Standing on the smooth sandy beach at the east end of the pond, in a calm September afternoon, when a slight haze makes the opposite shore-line indistinct, I have seen whence came the expression, "the glassy surface of a lake." When you invert your head, it looks like a thread of finest gossamer stretched across the valley, and gleaming against the distant pine woods, separating one stratum of the atmosphere from another. You would think that you could walk dry under it to the opposite hills you survey its surface critically, it is literally as smooth as glass, except where the skater insects, at equal intervals scattered over its whole extent, by their motions in the sun produce the finest imaginable sparkle on it, or, perchance, a duck plumes itself, or [...] a swallow skims so low as to touch it. (Thoreau 1982, 435)

Thoreau (1982, 366) himself earlier described "the glassy surface of the pond." Rather than a dead plane for the reflection of unholy nature, the pond's surface is living gossamer for the expression of the life of the pond and holy nature.

The "smooth reflecting surface" of ponds and lakes for Thoreau (1982, 339, 437) in *Walden* was "a perfect forest mirror" for connecting the earth below with the air above in what he twice called "sky water" (Thoreau

1982, 437, 444). With the addition of sunlight, reflective water is "a mirror in which all impurity presented to it sinks, swept and dusted by the sun's hazy brush." The sun is the master artist who paints in water-colours (literally) on the surface of lakes and ponds that, in turn, reflects the sky above. Thoreau (1962, *II*, 158; 1982, 437) wrote in his journal and in *Walden* that "a field of water betrays the spirit that is in the air. It is continually receiving new life and motion from above. It is intermediate in its nature between land and sky." Thoreau (1958, 186) also wrote in his journal that "water seems a middle element between earth and air." A swamp of water and trees, and a marsh of water and grass, are intermediate in their nature and a middle element between water and land. They portray the spirit that is in the water. They are continually receiving new life and motion from below, from what Benjamin calls in his neglected "Outline of the Psychophysical Problem" "the womb of the depths" (Benjamin 1996, 399) and in his essay on the tenth anniversary of Kafka's death "the dark, deep womb" (Benjamin 1973, 131; 1999b, 809). Benjamin and Thoreau found both of these wombs in swamps, vital facets of psychopolitical ecology.[4]

Living here at this time in the early nineteenth century and in this place in Massachusetts Thoreau achieves his desired self-transcendence and mystical union with holy nature:

> It is no dream of mine,
> To ornament a line;
> I cannot come nearer to God and Heaven
> Than I live to Walden even.
> I am its stony shore,
> And the breeze that passes o'er;
> In the hollow of my hand
> Are its water and its sand,
> And its deepest resort
> Lies high in my thought. (Thoreau 1982, 442; 2001, 26)

[4] As we will see with Thoreau in a later chapter. For Benjamin on swamps, see Giblett (2025, chapter 3).

Thoreau is united and consubstantial with the stony shore of Walden Pond expressed in a paradoxical mixture (or mishmash) of pagan Pantheism, Christian theology (or heresy), Romanticist pathetic fallacy, Transcendentalist mysticism (or hogwash), and Puritan plain-speaking composed in ornamental, pre-Romanticist Alexandrine rhyming couplets. All of these facets are compressed into a jewel of a short poem like a haiku that either encapsulates and sparkles with Thoreau's thought and feeling or annoys and irritates with his crotchets and contradictions. The water and sand of Walden Pond have been sublimated into the air above, the closest thing to God and Heaven for the young Transcendentalist and inwardly self-satisfied Thoreau. He is united with them, while connected to the depths of the pond.

Unlike the archetypal English Romanticist poet in William Wordsworth who related in a simile how he "wandered lonely as a cloud" in the heavens above the Lakes District,[5] Thoreau *is* the stony shore of the pond expressed in a metaphor of he who wandered not lonelily (as he was close to God) on the earth and water and in the heaven of his "lake country" (Thoreau 1982, 446). He engaged bodily with, and dreamt awake on, Walden Pond, unlike a Romanticist poet, such as Wordsworth did in dreamy poetry, and also related his experience in wide-awake prose. The modalities of poetry and prose are suited to their respective subjects (I/eye) and objects (water/lake/earth's eye/heaven).

Thoreau not only looked into the eye of a lake and his own depths but also knelt down on the ice of the pond and looked down into the depths of nature:

> Standing on the snow-covered plain, as if in a pasture amid the hills, I cut my way through a foot of snow, and then a foot of ice, and open a window under my feet, where, kneeling to drink, I look down into the quiet parlor of the fishes, pervaded by a softened light as through a window of ground glass, with its bright sanded floor the same as in summer; there a perennial waveless serenity reigns as in the amber twilight sky, corresponding to the cool and even temperament of the inhabitants. (Thoreau 1982, 525)

When the pond skimmed over with ice an inch thick Thoreau related in *Walden* that:

[5] For an ecocritical reading of Wordsworth's Romanticist landscape politics in *The Guide to the Lakes* and *The Prelude*, including of the country house in the gentleman's park estate, see Giblett (2011, 68–71, 80–82).

> The first ice is especially interesting and perfect, being hard, dark, and transparent, and affords the best opportunity that ever offers for examining the bottom where it is shallow; for you can lie at your length on ice only an inch thick, like a skater insect on the surface of the water, and study the bottom at your leisure, only two or three inches distant, like a picture behind a glass, and the water is necessarily always smooth then. (Thoreau 1982, 490–491)

Walden Pond grants access both to the watery depths below and to the airy heavens above. Not only "the universe is wider than our views of it," as Thoreau (1982, 559) puts it in the conclusion to *Walden*; it is also higher and deeper. When the surface of Walden Pond has thawed, heaven/the heavens above are reflected in its mirroring surface and Thoreau is the intermediatory eye/I between the two. When the surface of Walden Pond has frozen, heaven/the heavens above cannot be reflected in its mirroring surface. Heaven/the heavens are expressed instead in the glassy water in the depths below and Thoreau sees more deeply into it, himself, and the lives of others.

When the pond freezes over, it ceases to grant access to the heavens above by way of reflection, but grants access to the heavens below by means of mediation. Thoreau (1982, 524) related in *Walden* that "after a cold and snowy night it needed a divining-rod to find it. Every winter the liquid and trembling surface of the pond, which was so sensitive to every breath, and reflected every light and shadow, becomes solid to the depth of a foot or a foot and a half." Thoreau cuts through the ice a foot thick on the pond to open a window into the underlying water and its inhabitants. He concludes that "heaven is under our feet as well as over our heads" (Thoreau 1982, 525). For the later sacral and paludal Thoreau heaven and home are also in the swamps and meadows under and around our feet (as we will see in Chap. 9). Thoreau wrote in his journal and repeated in his essay "Autumnal Tints" that "heaven [or paradise] might be defined as the place which men avoid" (Thoreau 1962, *XI*, 127; 2001, 373). This certainly was the case with swamps (as we will also see in Chap. 9).

The depths of the pond is the home of the fishes, or more precisely their parlour, a living room, the same descriptor Thoreau uses for his own house on several occasions (as we saw in the previous chapter). He also relates walking home to his snug parlour from "some bright village parlor" (Thoreau 1982, 418) with its hard artificial light. The home of the fishes for Thoreau is a quiet pond parlour with its soft natural light. All three are living rooms with their "intimate charm"[6] for the gregarious and solitary naturalist who preferred his and the fishes' parlour to the village parlour.

Later in *Walden* Thoreau is scathing of the ice-cutter who "cuts and saws the solid pond, unroofs the house of fishes, and carts off their very element and air, held fast by chains and stakes like corded wood, through the favoring winter air, to wintry cellars, to underlie the summer there" (Thoreau 1982, 534–535). The ice-cutter has reduced the roof of the house of fishes to a resource to cool his drink in summer. Ice is like the timber cut and stacked to warm his house in winter. The living surface of the pond and the living trees of the wood have both suffered the same anthropocentric fate of becoming resources for human consumption and convenience without a thought for the living creatures for whom these places are home. The timber-cutter may utter oaths of admiration before chopping down a noble tree to cut it up into firewood, but does the ice-cutter utter the same oaths before chopping through a solid pond to cut it up into ice-blocks?

By the later stages of winter Thoreau becomes acutely aware and observant of the colour of water and its changing hues—frozen or thawed:

> Like the water, the Walden ice, seen near at hand, has a green tint, but at a distance is beautifully blue, and you can easily tell it from the white ice of the river, or the merely greenish ice of some ponds, a quarter of a mile off. [...] I have noticed that a portion of Walden which in the state of water was green will often, when frozen, appear from the same point of view blue. So the hollows about this pond will, sometimes, in the winter, be filled with a greenish water somewhat like its own, but the next day will have frozen blue. Perhaps the blue color of water and ice is due to the light and air they contain, and the most transparent is the bluest. Ice is an interesting subject for contemplation. [...When the pond has thawed] I shall look from the same window on the pure sea-green Walden water there, reflecting the clouds and the trees, and sending up its evaporations in solitude. [...] Perhaps I shall hear a solitary loon laugh as he dives and plumes himself, or shall see a lonely fisher in his boat, like a floating leaf, beholding his form reflected in the waves. (Thoreau 1982, 537–538)

[6] Paul Carter (1987, 261–292) discusses the English concept and practice of "intimate charm" in the Australian colonial context and experience.

Thoreau devoted a lengthy evocation in poetic prose earlier in *Walden* to a loon diving and laughing and his dialogic relationship with it (as we have seen in this chapter).

The inwardly self-satisfied Thoreau is becoming, or is also, the outwardly other-empathetic Thoreau who experiences and expresses fellow-feeling for fishes and their watery homes and who engages in social and economic critique. In the conclusion to *Walden* he also figures life in humans as rising and falling in sympathy and synchrony with the ebbs and flows of water and points to the fact that where he and his readers dwell today was wet land as "the life in us is like the water in the river. It may rise this year higher than man has ever known it, and flood the parched uplands; even this may be the eventful year, which will drown out all our muskrats. It was not always dry land where we dwell" (Thoreau 1982, 571). It was not always dry land, but wet land, where we dwell in cities today. Indeed, it was wetlands. It was not always dry land near where Thoreau dwelt next to Walden Pond, but wet land in Wyman Meadow. Indeed, it was and still is a wetland (as we will see in Chap. 9).

References

Arsić, Branka. 2016. *Bird Relics: Grief and Vitalism in Thoreau*. Cambridge, MA: Harvard University Press.

Bakhtin, M. M. 1981. *The Dialogic Imagination: Four Essays*. Edited by Michael Holquist. Translated by Caryl Emerson and Michael Holquist. Austin, TX: University of Texas Press.

Bakhtin, M. M. 1984. *Problems of Dostoevsky's Poetics*. Translated by Caryl Emerson. Minneapolis, MN: University of Minnesota Press. First published in Russian in 1963.

Bakhtin, M. M. 1986. *Speech Genres and Other Late Essays*. Edited by Caryl Emerson and Michael Holquist. Translated by Vern W. McGee. Austin, TX: University of Texas Press.

Benjamin, Walter. 1973. *Illuminations*. Translated by Harry Zohn. London: Fontana.

Benjamin, Walter. 1979. *One-Way Street and Other Writings*. Translated by Edmund Jephcott and Kingsley Shorter. London: NLB.

Benjamin, Walter. 1996. *Selected Writings, Volume 1: 1913–1926*, ed. Marcus Bullock and Michael Jennings. Cambridge, MA: Belknap Press of Harvard University Press.

Benjamin, Walter. 1999a. *The Arcades Project*. Translated by Howard Eiland. Cambridge, MA: Belknap Press of Harvard University Press.

Benjamin, Walter. 1999b. *Selected Writings, Volume 2: 1927–1934.* Edited by Michael Jennings, Howard Eiland and Gary Smith. Cambridge, MA: Belknap Press of Harvard University Press.

Benjamin, Walter. 2003. *Selected Writings: Volume 4, 1938–1940.* Edited by Howard Eiland and Michael Jennings. Cambridge, MA: Belknap Press of Harvard University Press.

Benjamin, Walter. 2019. *Origin of the German Trauerspiel.* Translated by Howard Eiland. Cambridge, MA: Harvard University Press.

Buell, Lawrence. 1995. *The Environmental Imagination: Thoreau, Nature Writing, and the Formation of American Culture.* Cambridge, MA: Harvard University Press.

Buell, Lawrence. 2024. *Henry David Thoreau: Thinking Disobediently.* New York: Oxford University Press.

Carter, Paul. 1987. *The Road to Botany Bay: An Essay in Spatial History.* London: Faber.

Giblett, Rod. 2011. *People and Places of Nature and Culture.* Bristol: Intellect Books.

Giblett, Rod. 2025. *Waking to Nature with Thoreau and Benjamin: Or, Psychopolitical Ecology.* New York: Bloomsbury.

Jefferies, Richard. 2001. In *At Home on the Earth: A New Selection of the Later Writings*, ed. Jeremy Hooker. Totnes: Green Books.

Jennings, Michael W. 1987. *Dialectical Images: Walter Benjamin's Theory of Literary Criticism.* Ithaca, NY: Cornell University Press.

Morson, Gary Saul, and Caryl Emerson. 1990. *Mikhail Bakhtin: Creation of a Prosaics.* Stanford: Stanford University Press.

Thoreau, Henry David. 1958. *Consciousness in Concord: The Text of Thoreau's Hitherto 'Lost Journal' (1840–41), Together with Notes and a Commentary by Perry Miller.* Boston: Houghton Mifflin.

Thoreau, Henry David. 1962. *The Journal 1837–1861, Volumes I–XIV.* Edited by Bradford Torrey and Francis Allen. New York: Dover.

Thoreau, Henry David. 1982. *The Portable Thoreau.* Edited by Carl Bode. New York: Viking Penguin.

Thoreau, Henry David. 2001. *Collected Essays and Poems.* New York: Library of America.

Wood, David F. 2006. *An Observant Eye: The Thoreau Collection at the Concord Museum.* Concord, MA: Concord Museum.

CHAPTER 6

Nature Writing Is Multi-Generic and Heterogenic

Nature writing is multi-generic and heterogeneous in letter, memoir, essay, story, and journal. Nature writing creates accounts in these genres about nature at home or away with either going on excursions by foot or boat to exotic places away from home in search of people, plants, and animals or sauntering in his home place in appreciation of local places, people, plants, and animals. Thoreau did both. White created the genre of nature writing about the home place, its plants, and animals; Thoreau perfected it about home and away.

Genre for Bakhtin is the primary form of utterance whether it be in speech or writing, or any other media for that matter. For him "we speak only in definite speech genres, that is, all our utterances have definite and relatively stable typical *forms of construction of the whole*" (Bakhtin 1986, 78; his emphases). Speech genres for Bakhtin include all other types of utterance. In addition for him "to learn to speak means to learn to construct utterances" (Bakhtin 1986, 78). They are formed into speech genres. To learn to speak is also to learn genres. No speech without genres.

6.1 Nature Writing Is Multi-generic

Nature writing mixes genres. As Jacques Derrida said in "The Law of Genre," "you cannot not mix genres." The law of genre is to mix genres. All genres are mixed, the mixture of other, previous genres. Genres also

R. Giblett, *The Politics of Nature and Environmental Writing*, Transdisciplinary Environmental Humanities 1,
https://doi.org/10.1007/978-3-032-04427-3_6

"*grow together, inosculate*, or *knit together* (as bones are said to knit together)" (Morson and Emerson 1990, 293; their emphases). Genres are living beings. Genres "carry and shape social experience for individual people" (Morson and Emerson 1990, 271). "The work" for Morson and Emerson (1990, 272) "is a social fact oriented toward an audience." The work is also a generic fact oriented towards the past. Genre for Bakhtin is "a key organ of memory and important vehicle for historicity" (Morson and Emerson 1990, 280). Genres are born, live, die, and are reborn in new mixtures. "The life of a genre" for Bakhtin (1984, 141) "consists in its constant rebirths and renewals in *original* works." Nature writing is exemplary in this regard as it has been reborn and renewed in original works since its birth in White's *Selborne*.

For Bakhtin:

> [A] newly born genre never supplants or replaces any already existing genres. Each new genre merely supplements the old ones, merely widens the circle of already existing genres. [...] Thus no new artistic genre ever nullifies or replaces old ones. But at the same time each fundamentally and significantly new genre, once it arrives, exerts influence on the entire circle of old genres. (Bakhtin 1984, 271)

These are certainly the case on every count with nature writing. The old genres of letter, memoir, essay, story, and journal live on in it. Nature writing embraced them in a wider circle and exerted an influence on them to bring new topics into their narrower circle.

A genre for Bakhtin (1984, 106) "lives in the present, but always remembers its past, its beginning. Genre is a representative of creative memory." Memory is never completely accurate recall of the past. A genre is creative memory of its beginning in past genres. Genres for Bakhtin (1986, 65) are "the drive belts from the history of society to the history of language" and literature. Thoreau's *Walden* and "Walking" are exemplary in this regard, driving the belt connecting the history of American society from its beginnings with dispossessed indigenous people to the history of the English language and literature from Chaucer to Coleridge.

"The creation of new genres" for Morson and Emerson (1990, 276; see also 278, 281, 371), such as the novel and nature writing, comes about because the new genre was needed to do work that the old genres did not and could not do. "Artistic form" for Bakhtin (1984, 43) "does not shape already prepared and found content, but rather permits content to be found and seen for the first time." Genres for Morson and Emerson (1990, 282) "convey a vision of the world not by explicating a set of propositions but by developing concrete examples." Such was the case with Thoreau throughout his writing career, beginning with *A Week on the Concord and Merrimack Rivers.*

The title of *A Week on the Concord and Merrimack Rivers* suggests for Robert Thorson (2017, 8) "a travel narrative," but for him the title is misleading as "the bulk of the book is a scattered anthology of transcendental musings, "doggerel verse," natural history description, events unrelated to the fictional chronology, which distils two weeks into one." One could say the same thing about *Walden* (except that it distils two years into one), or that it is about Thoreau building a house by a pond and hoeing beans in a nearby field he cleared of weeds. This is like saying Homer's *Odyssey* or Melville's *Moby Dick* is book about a man who goes on a sea voyage. Similarly *A Week* is not just about the river journey Thoreau and his brother took over two weeks in a boat he and his brother built in a week. Nor is it just, as Thorson (2017, 93) puts it later, "a random collection of scholarly essays, poems, and digression." One could also say the same thing about *Walden,* which is for Thorson (2017, xiii) "arguably the most important work of American nonfiction." *A Week* is arguably the fourth most important work of American non-fiction after Rachel Carson's *Silent Spring* and Aldo Leopold's *A Sand County Almanac.* Like *Walden,* *A Week* weaves together nature writing, family and environmental history, mythology, philosophy, religion, and science. John McPhee (2004, xxviii) describes Thoreau's writing in *A Week* as "commentary, editorial, philosophical, homiletic," as "defying generic assignment" and as "between psychiatry and religion," all apt descriptors, and for *Walden* too.

Bakhtin (1984, 108) outlines the characteristics what he calls "serio-comical genres" one of which is that they are:

> a multi-toned narration, the mixing of high and low, serious and comic; they make wide use of inserted genres—letters, found manuscripts, retold dialogues, parodies on the high genres, parodically reinterpreted citations; in some of them we observe a mixing of prosaic and poetic speech, living dialects and jargons [...] are introduced, and various authorial masks make their appearance.

Walden is serio-comical nature writing. It does most of the things that Bakhtin mentions, including retelling dialogues in prosaic speech between

Thoreau and local identities, such as the woodchopper. It includes some of Thoreau's own poems in poetic speech. It parodies the high genre of the epic in his long, mock-epic story of the battle between the red and black ants (Thoreau 1982, 474–478).

Benjamin begins his essay "The Image of Proust" by stating that "[t]he thirteen volumes of Marcel Proust's *À la Recherche du Temps Perdu* are the result of an unconstruable synthesis in which the absorption of a mystic, the art of a prose writer, the verve of a satirist, the erudition of a scholar, and the self-consciousness of a monomaniac have combined in an autobiographical work" (Benjamin 1973, 203; 1999, 237). Benjamin's lifelong friend Gershom Scholem rejigged this pronouncement and applied it to Benjamin in the dedication to him of *Major Trends in Jewish Mysticism*, excluding "monomaniac" (cited by Friedlander 2012, 223 n1). Precisely the same things on all counts could be said, and many of them have been said, about the fourteen volumes of Thoreau's *Journal* and his other works, especially "the unconstruable synthesis" of *Walden*, an apt three-word descriptor.

Benjamin (1973, 203; 1999, 237) went on in his essay on Proust to elaborate further in terms that are also applicable to *Walden* that "it has been rightly said that all great works of literature found a new genre or dissolve one—that they are, in other words, special cases. Among these cases this [Proust's *À la Recherche du Temps Perdu*] is one of the most unfathomable." Benjamin is here re-purposing a line of thought from his study of the German mourning play, in which he argued that "the important works [...] stand outside the boundaries of genre. An important work either founds the genre or dissolves it; and in perfect works the two functions unite" (Benjamin 2019, 23). Benjamin did not give to *À la Recherche du Temps Perdu* the accolade of the perfect work for uniting the two functions of foundation and dissolution of a genre. *Walden* is certainly an important work and a special case standing outside the boundaries of genre (as Thoreau aimed to do as we have seen). The same things could also be said about *The Arcades Project.* Arguably *Walden* is also a perfect work as it founded the genre of American nature writing and, at the same time and in the same place, dissolved it into wild writing, memoir, satire, and mystical saint's life, as well as into a wild loon chase (as we have seen) and a wild goose chase (as we will see in a later chapter).

Benjamin (1973, 203; 1999, 237) went on in his essay on Proust to argue that *À la Recherche du Temps Perdu* "from its structure, which is fiction, autobiography, and commentary in one, to the syntax of endless sentences (the Nile of language, which here overflows and fructifies the regions of truth), everything transcends the norm." The same thing could be said about *Walden*, and many of them have been said, with the proviso that some of its sentences are not quite as long as some of those of *À la Recherche du Temps Perdu*. Thoreau would surely have appreciated Benjamin's figure of the unfathomable as he wrote in *A Week* that "the shallowest still water is unfathomable" (Thoreau 1998, 39). He also explored in *Walden* and his other writings the unfathomable depths of himself in the water of Walden Pond and in the shallowest still water of the swamps of Concord. He explored the unfathomable depths of the grotesque lower bodily, earthly, and mental strata. He would have also appreciated Benjamin's figures of the Nile of language as he wrote about the Concord and Merrimack Rivers in *A Week* that overflowed and fructified the regions of truth in *Walden* and his other writings. *Walden* upsets the norms, including the norms of genres, such as landscape writing and the idyll.

6.2 Landscape Writing and the Idyll 1

Walden is nature writing, not landscape writing, as "'landscape'," Bakhtin (1981, 143–144; his scare quotes), argues:

> is born [when] nature [is] conceived as horizon (what a man sees) and as the environment (the background, the background, the setting) for a completely private, singular individual who does not interact with it. Nature of this kind differs sharply from nature conceived in a pastoral idyll or georgic.

The pastoral idyll and georgic were two of Thoreau's favourite genres and its writers particularly in Cato, Varro, and Virgil were among his favourites. Cato and Varro wrote idyllic guides to agriculture and horticulture; Virgil wrote ecologic and georgic celebrations of living in the country. Like the classical epic for Bakhtin (1981, 218), in pastoral idyll and georgic "there is no landscape, no immobile dead background; everything acts, everything takes part in the unified life of the whole." The same could be said of Thoreau's nature writing.

Bakhtin (1981, 225) defines the idyll by "the unity of place" and "the special relationship that time has to space in the idyll." In the idyll for Bakhtin there is:

> an organic fastening down, a grafting of life and its events to a place, to a familiar territory with all its nooks and crannies, its familiar mountains, valleys, fields, rivers, and forests, and one's own home. (Bakhtin 1981, 225)

This is the stock-in-trade of nature writing from White through Thoreau to 'the new nature writing' and environmental writing (as we will see in later chapters). In case of the Roman idyll writers, ponds could be added to the list as they were good for domesticated fowl. In Thoreau's case ponds could be added to the list too as they were good for birds, fish, and self-reflection (Giblett 2025, chapter 1). Mountains, valleys, fields, rivers, and forests, with the addition of lakes, were also privileged by the Romanticist idyll writers, such as Wordsworth (Giblett 2011, chapter 3). Thoreau inherited the Roman and Romanticist traditions and reproduced them uncritically in his early writings, such as *A Week* and *Walden*. He also felt embarrassed that his woods were not a forest and his pond was not a lake (though he called it as such occasionally). The scenery of Walden was on a modest scale and not as grandiose as the English Lakes District was for Wordsworth.

In Thoreau's case and in his later writings, such as "Walking," he added marshes and swamps to the list of idyllic places. Wetlands, such as marshes, swamps, and tarns, were not idyllic places for Cato, Varro, Virgil, and Wordsworth, all of whom denigrated them (Giblett 2011, chapter 3; 2025, chapter 5). Thoreau portrayed swamps as idyllic sacred places (as we will see in a later chapter). He subverted the hierarchical privileging of mountains, valleys, fields, rivers, lakes, ponds, woods, forests, and cities over swamps and marshes in the Roman and Romanticist idylls by portraying marshes and swamps as idyllic and preferring them to the idyllic geographical features.

The swamp and marsh, though, were not his home place where Thoreau could live and did live. They retained their uncanniness and unfamiliarity with their sacrality. The familiar territory is one's own home place, in Thoreau's case by the pond in the woods during his two-year sojourn in his house there. "Idyllic life and its events" for Bakhtin (1981, 225) are "inseparable from this concrete, spatial corner of the world" of one's own home place, as they were for Thoreau (1962, *V*, 254) for whom "home is here." "This little spatial world" for Bakhtin (1981, 225) is "limited and sufficient unto itself, not linked in any intrinsic way with other places, with the rest of the world." Yet this was not the case for Thoreau who linked his home place with other places in America and the rest of the world.

The idyll for Bakhtin has a special relationship not only to place and space, but also to time coalescing in what he calls "a chronotope," "literally, 'time space'" (Morson and Emerson 1990, 367). A chronotope is a way of thinking together "actions and events" in time and space (Morson and Emerson 1990, 367). The idyll's unity of place for Bakhtin (1981, 225) "also contributes in an essential way to the creation of the cyclic rhythmicalness of time so characteristic of the idyll." The narrative genre of the idyll for Bakhtin is a rich chronotope for presenting the cycles of time of the sun, moon, days, months, years, and the seasons. Writing about the seasons has endured for four centuries from James Thomson in the eighteenth century, through Thoreau in the nineteenth century and Leopold in *A Sand County Almanac* in the twentieth century, to the twenty-first century with Rick Bass in *Wild Marsh: Four Seasons at Home in Montana* (Bass 2009) and *Black Swan Lake: Life of a Wetland* (Giblett 2013) based on the six seasons of the local Aboriginal people.

6.3 Writing the Book of the Seasons

Thoreau bemoaned early on in his journal that "no one, to my knowledge," had written "a book of the seasons [...] in its own season and out-of-doors, or in its own locality" (1962, *II*, 239). Thoreau is as good as his word as he did all that and more in *Walden*. It is now what Buell (1995, 232) calls "the most famous of all American season books." It is much more than that though, even when it comes to the seasons. Thoreau encourages the readers of his essay on huckleberries, repeated from his journal, to "live in each season as its passes; breathe the air, drink the drink, taste the fruit, and resign yourself to the influences of each. Let these be your only diet-drink and botanical medicine" (1962, *V*, 394; 2001, 501). In other words, eat, drink, and breathe locally, merrily, and seasonally, not globally and trans-seasonally. Later he wrote in his journal "I love best to have each thing in its season only, and enjoy doing without it all other times" (Thoreau 1962, *IX*, 160).

One way in which Thoreau (2001, 386) in his essay "Autumnal Tints" advocated an appreciation for the seasons was by having street trees planted that "mark the season. [...] Let us have Willows for spring, Elms for summer, Maples and Walnuts and Tupeloes for autumn, Evergreens for winter and Oaks for all seasons." Australians should mark the Aboriginal seasons of their local place by appreciating the native species that flower in succession, often through six months of the year in some highly biodiverse places.

Living by Walden Pond and Wyman Meadow, Thoreau was awake to the seasons, initially to spring, later to winter and autumn (fall). He wrote in *Walden* that "one attraction in coming to the woods to live was that I should have leisure and opportunity to see the Spring come in" (Thoreau 1982, 542). It was not an attraction that kept him there for long, nor a reason for staying there, beyond a couple of springs as he left after two years, two months, and two days (Thoreau 1982, 258). Seeing the spring come in is time- and place-dependent. In the woods it is more visible, more pronounced. Being awake to nature's seasons is a lifelong awakeness.

Thoreau devoted an entire chapter of *Walden* to spring in which he writes:

> In almost all climes the tortoise and the frog are among the precursors and heralds of this season, and birds fly with song and glancing plumage, and plants spring and bloom, and winds blow, to correct this slight oscillation of the poles and preserve the equilibrium of nature. As every season seems best to us in its turn, so the coming in of spring is like the creation of Cosmos out of Chaos and the realization of the Golden Age. (Thoreau 1982, 553)

Every season has its time and place in the cycle of growth, decay, and regrowth. After the dormancy and hibernation of winter, comes the divine creation of spring. Winter is pregnant with spring. Thoreau (1982, 510) "rarely failed to find, even in midwinter, some warm and springy swamp where the grass and the skunk-cabbage still put forth with perennial verdure, and some hardier bird occasionally awaited the return of spring." Including humans. Thoreau (1982, 551) also wrote in his chapter on spring that "so our human life but dies down to its root, and still puts forth its green blade to eternity." It is in synch with other life, including in Thoreau's case Walden Pond that "was dead and is alive again" (Thoreau 1982, 551). Thoreau is taking God's words from Luke 15: 24 and 32 about the death and resurrection of Jesus. For Thoreau, Walden is resurrected as a latter day Jesus, the son and word of God.

When Thoreau went to the woods to see spring, he also saw winter and appreciated it too:

> In the deepest snows, the path which I used from the highway to my house, about half a mile long, might have been represented by a meandering dotted

line, with wide intervals between the dots. For a week of even weather I took exactly the same number of steps, and of the same length, coming and going, stepping deliberately and with the precision of a pair of dividers in my own deep tracks—to such routine the winter reduces us—yet often they were filled with heaven's own blue. (Thoreau 1982, 508)

In winter Thoreau could not see heaven's blue reflected in the water of the pond, but could see it in his footprints.

Heaven's blue is also the colour of the robed visitor Thoreau had in his last winter at the pond. He wrote in *Walden*:

> I should not forget that during my last winter at the pond there was another welcome visitor. [...] One of the last of the philosophers. [...] He is perhaps the sanest man and has the fewest crotchets of any I chance to know; the same yesterday and tomorrow. Of yore we had sauntered and talked, and effectually put the world behind us; for he was pledged to no institution in it, freeborn, *ingenuus*. Whichever way we turned, it seemed that the heavens and the earth had met together, since he enhanced the beauty of the landscape. A blue-robed man, whose fittest roof is the overarching sky which reflects his serenity. I do not see how he can ever die; Nature cannot spare him. (Thoreau 1982, 511–512)

This blue-robed wizard is a personification of the union of the heavens and the earth, a god who can never die.

Thoreau counters the traditional tyrannical personification of winter and proposes a more benign personification that harks back to the previous summer. He wrote in his early essay "A Winter Walk":

> Though winter is represented in the almanac as an old man, facing the wind and sleet, and drawing in his cloak around him, we rather think of him as a merry woodchopper, and warm-blooded youth, as blithe as summer. (Thoreau 1982, 74)

He wrote in *Walden* along similar lines how:

> Many of the phenomena of Winter are suggestive of an inexpressible tenderness and fragile delicacy. We are accustomed to hear this king described as a rude and boisterous tyrant; but with the gentleness of a lover he adorns the tresses of Summer. (Thoreau 1982, 550)

Thoreau was not a proponent of a rigid distinction between and within the seasons. He wrote in his journal that "the year has many season more than a recognized in the almanac" (Thoreau 1962, *II*, 21). He described in "A Winter Walk" how in winter "Nature confounds her summer distinctions at this season. The heavens seem to be closer to the earth. The elements are less reserved and distinct. Water turns to ice, rain to snow" (Thoreau 1982, 63).

In winter, the elements blur and transform. Liquid water becomes solid ice, through semi-liquid and -solid snow; in spring and summer, solid ice becomes liquid water under the heat of the sun.

After summer comes autumn (or fall), to whose tints Thoreau devoted an entire essay. One of the attractions for Thoreau going to the woods was to see spring come in, but he also saw autumn while living there and saw more than heaven and himself reflected in the pond. He wrote in *Walden* that "it is a soothing employment, on one of those fine days in the fall when all the warmth of the sun is fully appreciated, to sit on a stump on such a height as this, overlooking the pond, and study the dimpling circles which are incessantly inscribed on its otherwise invisible surface amid the reflected skies and trees" (Thoreau 1982, 436).

Autumn is the time of the changing colours of foliage, though these are not as dramatic in England as they are in New England. Thoreau (2001, 367) begins his essay "Autumnal Tints" by noting:

> Europeans coming to America are surprised by the brilliancy of our autumnal foliage. There is no account of such a phenomenon in English poetry, because the trees acquire but few bright colors there. The most that [James] Thomson says on this subject in his "Autumn" [in his collection of long poems, *The Seasons*] is contained in the lines—
>
> 'But see the fading, many-coloured woods,
> Shade deepening over shade, the country round
> Imbrown; a crowded umbrage, dusk and dun,
> Of every hue from wan declining green/ To sooty dark':—[...]
>
> The autumnal change of our woods has not made a deep impression on our own literature yet. October has hardly tinged our poetry.

Thoreau's nature writing quotes the poetry of other poets. Thoreau goes on to rectify this situation in the remainder of his essay to show how American autumnal foliage has made a deep impression on him and his senses, and a first impression on American literature beginning with him, and this essay.

For Thoreau there are seasons, not only of the year, but also of the day, history, and the soul. The day is four seasons compressed into one, not just that it can be hot and cold, wet and dry in the same day (as is the case with the city of Melbourne where I live that famously can have four seasons sometimes in one day). For Thoreau (1982, 541) "the day is an epitome of the year. The night is the winter, the morning and evening are the spring and fall, and the noon is the summer." The morning for Thoreau (1982, 542) is "the most memorable season of the day." Morning and spring for Thoreau are "the awakening of nature" (Thoreau 1962, *XI*, 455). What better time for waking to nature when nature is awakening.

For Thoreau there are also seasons of history. While discussing Geoffrey Chaucer, the author of *The Canterbury Tales* and the chronicler of early modern English, Thoreau (2001, 145) associates the seasons with the ages of literary history: "our summer of English poetry is well-advanced towards its fall, and laden with the fruit and foliage of the season, with bright autumnal tints, but soon the winter will scatter its myriad clustering and shading leaves, and leave only a few desolate and fibrous boughs to sustain the snow and rime, and creak in the blasts of ages." Naturally he places himself in the flowering and fruitful season of summer and prophesises the fall and frosty winter (of modernism?) to come.

For Thoreau, like John Keats for whom there are "seasons in the mind of man," there are seasons in the landscape of the mind. He writes in his essay "Walking" about how "in some more genial season, perchance, a faint shadow flits across the landscape of the mind, cast by the *wings* of some thought in its vernal or autumnal migration" (Thoreau 1982, 627; his emphasis). He advised himself and his readers who are would-be writers that "it is in vain [for you] to write on the seasons unless you have the seasons in you" (Thoreau 1962, *X*, 253). Thoreau had followed his own advice as he wrote earlier in his journal that "the seasons and all their changes are in me" (Thoreau 1962, *X*, 127), so he was well-placed to give advice.

For Thoreau there is also a season of the soul as he relates in his poem "The Soul's Season":

> A sober mind will walk alone,
>   Apart from nature if need be,
> And only its own seasons own,
>   For nature having its humanity.

The season of the soul is not necessarily in synch with the season of the earth and sky for:

Sometimes a late autumnal thought
  Has crossed my mind in green July,
And to its early freshness brought
  Late ripened fruits and an autumnal sky. (Thoreau 2001, 560, 561)

A mature, autumnal thought crosses Thoreau's mind belatedly in green, youthful summer and changes the face of summer into autumn. Conversely what he calls "a dry but golden" autumnal thought crosses prematurely the summer greenness of his mind and makes mature autumnal wisdom linger into green and youthful summer:

A dry but golden thought which gleamed
  Athwart the greenness of my mind,
And prematurely wise it seemed,
    Too ripe "mid summer's youthful bowers to find.
(Thoreau 2001, 560, 561)

The season of the soul may be out of synch with the season of the earth, yet it is still the season of autumn. He proclaims in another poem without a title that "I am the Autumnal sun" and later that "the winter is lurking within my moods" (Thoreau 2001, 583). Thoreau is in no doubt that there are seasons of the soul. He is also drawing on a long history of associating the seasons with age, but he upsets the traditional association by dissociating the seasons of the soul, or mind, with the seasons of the earth. Thoreau deconstructs and decolonizes the European cultural construction of the seasons, and the second, urban/agricultural culture of nature. [1]

The cyclic rhythmicalness of time also manifests in the "idyllic-cyclical" chronotope of the *Bildungsroman*, or novel of emergence (Bakhtin 1986, 21–22), "the only chronotope that assimilates real historical time" for Bakhtin (Morson and Emerson 1990, 413). Yet this is also the case with

[1] For further discussion of the cultures of natures, see Giblett (2011). For further discussion of the seasons drawing on Geoffrey Chaucer, James Thomson, and others, see Giblett (2013, chapter 18).

the 'non-fiction' *Bildungsroman* of *Walden* and Thoreau's journal with their memoirs of the life, education, and emergence of the young Thoreau that assimilates the real historical time of himself and his age (in two senses). They also assimilate a real geographical place in space. The features and locales of this place have been assimilated by being inscribed in maps reproduced in many editions of Thoreau's books, studies of his work, and in a poster map (Walden Woods Project 1991). They present only the spatial aspect of the chronotope of Thoreau's *Walden*, whereas Thoreau could not and did not separate space and time in *Walden*; they were intertwined in life and writing. Maps freeze a moment in time; they are synchronic and not diachronic. They cannot and do not show, for instance, the rising and falling water levels of ponds and swamps over time. Chronotopes are synchronic and diachronic. They are temporally simultaneous and sequential, and spatially co-existent and extensive.

The chronotope for Bakhtin (1986, 34) takes "a given locality [and] transforms a portion of terrestrial space into a place of historical life." This is an apt descriptor for *Walden* that takes the given localities and the portion of terrestrial space of Walden Woods and Pond and transforms them into a place of the historical life of the 1820s to 1850s. The chronotope for Bakhtin (1986, 41–42) is the merging of the past with present, the connection of the present and the past, and the traces and presence of the past in the present that "pervades localized time," including the future. *Walden* does this when Thoreau (1982, 272) wrote in it that he stood "on the meeting of two eternities, the past and future, which is precisely the present moment." In other words, he was located spatially and temporally in a chronotope.

6.4 Landscape Writing and the Idyll 2

Unlike landscape writing, the idyll is characterized by what Bakhtin (1981, 226) calls "the conjoining of human life with the life of nature," as in *Walden* and *A Sand County Almanac*. This conjunction is especially the case with what Bakhtin (1981, 227) goes on to call "the labor aspect of this [agricultural] idyll" in which "the agricultural-labor element [...] creates a real link and common bond between the phenomena of nature and the events of human life." In Leopold's case, fixing up a chicken shack into a habitable house and planting endemic species of trees on a rundown farm required and demonstrated both architectural and environmental labour. In Thoreau's case, hoeing a field of beans to grow them for sale and building his own house to live in required and demonstrated both agricultural and architectural labour. He earnt his living by the labour of his hands for the two years he lived by Walden Pond, as he said (Thoreau 1982, 258).

In the agricultural idyll Bakhtin (1981, 227) describes how:

> people consume the produce of their own labor, the produce is figurally linked with the productive process, in it—in this produce—the sun, the earth and the rain are actually present (not merely in some system of metaphorical links).

Even though Thoreau did not consume his beans (he tasted them) and sold them, the beans were the product of his work and they were physically, materially, and figurally linked with the productive process of hoeing in which the sun, the earth, and the rain are actually present. Other actors were present too, such as weeds that he cut down in heroic mock-epic fashion (Thoreau 1982, 406, 411) and woodchucks "nibbling off the earliest tender leaves" (Thoreau 1982, 413). Thoreau performed what Bakhtin calls "the unmechanized nature of idyllic labour," whereas Cato, Varro, and Virgil had slaves and overseers to do the work, a fact of their idylls that they mention which was common for their time and place, but Bakhtin neglects to mention it. Bakhtin had an idyllic view of the idyll.

The idyll is problematic in recent nature writing about a colonized place in which the history of the past, including a massacre, with Nicholas Shakespeare's *In Tasmania*. The tension or cross-fertilization between travel writing and nature writing, and between memoir and essay in nature writing, is also brought into stark relief in *In Tasmania*. Following in the footsteps of Bruce Chatwin and his *In Patagonia*, Shakespeare travelled to the exotic place of Tasmania, not on an excursion like Chatwin to Patagonia, but to live. Shakespeare interweaves the history of Tasmania and his family history with the account of his sea-change moving to, and living near, Moulting Lagoon on the central east-coast of the island-state. He is an outsider and 'newbie' who seems to have appreciated the idyllic backwater of Moulting Lagoon on the front doorstep of Tasmania more than native Tasmanian writers. He is an ex-pat Pom and noted writer on the byways of English history and writing, including books on Ian Fleming and Bruce Chatwin. *In Tasmania* was first published in 2004 with the subtitle of *Adventures at the End of the World* as a kind of 'golly-gosh' book about an exotic place much-loved by armchair tourist Brits. Tasmania for many is a good place spatially at the end of the world to observe the end of the world temporally in a chronotope.

Shakespeare relates an Aboriginal story about the creation of the black swan. He describes he and his son finding a dying black swan on the beach near where they live that "must have been heading for Moulting Lagoon." This encounter prompts him to recall "the story of the black swan [which] was the most beautiful of the legends that Tapte, chief of the lagoon people, passed on to Timler before his death in 1820." The legend of the black swan tells of the creation of Pickerdas, the black swan, from Mitaweena, the dying whale, who:

> blew spouts of black mud into the air from his blowhole. "In his last struggles the muddy jets flew higher and higher until reaching the rarer atmosphere, they turned into great black birds wheeling and trumpeting above the body of their dying creator. They were as black as the mud from Mitaweena's grave and their eyes and beaks were blood red from the bursting lungs of the great whale. Their wings were tipped by the clouds as they soared high above his last resting place in a sad salute of farewell. And this was the coming of Pickerdas the black swan." (Shakespeare 2007, ebook)

Shakespeare is citing this legend from Jackson Cotton's collection of his and his grandparents' recollection of Aboriginal Tasmanian legends, *Touch the Morning*, first published in 1979 and republished in 2013 as *The Land of the Sleeping Gods* with an historical introduction and notes, as well as a prologue by the noted Australian historian Henry Reynolds.

Shakespeare also reiterates that Moulting Lagoon was "a vital gathering place for the Oyster Bay Tribe." It was fenced off by the colonist George Meredith depriving the tribe of food. It was also the site of what Shakespeare calls "not only an alleged Aboriginal massacre but of a slaughter of thylacines" (also known as 'Tasmanian tigers' because of their striped coat, now extinct). Moulting Lagoon has a mixed history as a breeding site for new life, a shedding site for the recreation of life, and a killing site for massacre and slaughter. *In Tasmania* is mega multi-generic comprising travel writing, nature writing, memoir, and essay on local, Tasmanian, and family history. Shakespeare interweaves Aboriginal story of people, places, and a species with colonial history in postcolonial times.

References

Bakhtin, M. M. 1981. *The Dialogic Imagination: Four Essays*. Edited by Michael Holquist. Translated by Caryl Emerson and Michael Holquist. Austin, TX: University of Texas Press.

Bakhtin, M. M. 1984. *Problems of Dostoevsky's Poetics*. Edited and translated by Caryl Emerson. Minneapolis, MN: University of Minnesota Press. First published in Russian in 1963.

Bakhtin, M. M. 1986. *Speech Genres and Other Late Essays*. Edited by Caryl Emerson and Michael Holquist. Translated by Vern W. McGee. Austin, TX: University of Texas Press.

Bass, Rick. 2009. *Bass in Wild Marsh: Four Seasons at Home in Montana*. Boston: Houghton Mifflin Harcourt.

Benjamin, Walter. 1973. *Illuminations*. Translated by Harry Zohn. London: Fontana.

Benjamin, Walter. 1999. *Selected Writings, Volume 2: 1927–1934*. Edited by Michael Jennings, Howard Eiland and Gary Smith. Cambridge, MA: Belknap Press of Harvard University Press.

Benjamin, Walter. 2019. *Origin of the German Trauerspiel*. Translated by Howard Eiland. Cambridge, MA: Harvard University Press.

Buell, Lawrence. 1995. *The Environmental Imagination: Thoreau, Nature Writing, and the Formation of American Culture*. Cambridge, MA: Harvard University Press.

Friedlander, Eli. 2012. *Walter Benjamin: A Philosophical Portrait*. Cambridge, MA: Harvard University Press.

Giblett, Rod. 2011. *People and Places of Nature and Culture*. Bristol: Intellect Books.

Giblett, Rod. 2013. *Black Swan Lake: Life of a Wetland*. Bristol: Intellect Books.

Giblett, Rod. 2025. *Waking to Nature with Thoreau and Benjamin; Or, Psychopolitical Ecology*. New York: Bloomsbury.

McPhee, John. 2004. Introduction. In *A Week on the Concord and Merrimack Rivers*, ed. Henry David Thoreau, ix–xlvi. Princeton, NJ: Princeton University Press.

Morson, Gary Saul, and Caryl Emerson. 1990. *Mikhail Bakhtin: Creation of a Prosaics*. Stanford: Stanford University Press.

Shakespeare, Nicholas. 2007. *In Tasmania*. North Ryde: Random House Australia.

Thoreau, Henry David. 1962. *The Journal 1837–1861, Volumes I–XIV*. Edited by Bradford Torrey and Francis Allen. New York: Dover.

Thoreau, Henry David. 1982. *The Portable Thoreau*. Edited by Carl Bode. New York: Viking Penguin.

Thoreau, Henry David. 1998. In *A Week on the Concord and Merrimack Rivers*, ed. H. Daniel Peck. New York: Penguin. First published in 1849.

Thoreau, Henry David. 2001. *Collected Essays and Poems*. New York: Library of America.

Thorson, Robert. 2017. *The Boatman: Henry David Thoreau's River Years*. Cambridge, MA: Harvard University Press.

Walden Woods Project, The. 1991. *Walden Woods: Cradle of American Conservation*. Boston: The Walden Woods Project.

CHAPTER 7

Nature Writing Is Carnivalesque About the Grotesque and the Monstrous

Thoreau was less a tame nature writer and more of a wild nature writer. He was an early advocate for immersion in nature, especially in what he called "wildness" by which he did not just mean the wilderness without, but also the wildness within, "the bog in our brains and our bowels," as he put it in his journal (Thoreau 1962, *IX*, 43). Wildness is the grotesque lower strata of the body and the mind (as we will see later in the present chapter), as well as the earth, especially swamps (as we will see in a later chapter). The grotesque lower strata of the body, the mind, and the earth are where monsters, such as alligators, live (as we will see later in the present chapter). The monstrous lives in the grotesque lower strata. The grotesque lower strata of the body, the mind, and the earth come together in psychopolitical ecology.

The grotesque body is the belching, farting, gluttonous, lusting, creative, procreative body; it is a body having sex, giving birth, being born, eating and shitting, living and dying; it is the body of what Bakhtin (1984, 9) called "the popular sphere of the marketplace" (as distinct from "the official sphere" of the market) and it is the popular body (as distinct from the official body). Bakhtin was the champion of the grotesque body and the carnivalesque in his book on Rabelais first published in Russian in 1965 and later translated into English (Bakhtin 1984). The grotesque body for Bakhtin was a vehicle and vector for the carnivalesque in which the hierarchies of norm society were subverted (Bakhtin 1984, 10, 15, 81,

R. Giblett, *The Politics of Nature and Environmental Writing*, Transdisciplinary Environmental Humanities 1,
https://doi.org/10.1007/978-3-032-04427-3_7

285–286; Morson and Emerson 1990, 435–439, 444, 449–450). The carnivalesque for Bakhtin (1984, 123) "brings together, unifies, weds, and combines the sacred with the profane." Thoreau's *Walden* and "Walking" do likewise. Both also celebrate "the reproductive power of the earth and the body" in the grotesque body. Thoreau was a prankster of the carnivalesque who subverted the hierarchies of tame over wild, brain over bowels, city over swamp. The carnivalesque manifests the grotesque body, mind, and places as a whole. The grotesque for Bakhtin (1984, 19–20) unified indissolubly "the sphere of earth and body" in the body of the earth, especially in the lower strata of the nether regions of both.

Allon White (1993, 167) defines the grotesque as the "perilous sumps of ambiguity." The grotesque is the place in a culture to which the discarded misfits of meaning run to wallow in a mire of monstrosity. The grotesque is a dangerous swamp of monstrosity (Bakhtin 1984, 43; Giblett 1996, chapter 8). Grotesque places are not only geographical places but also corporeal places. Grotesque places are physical places, both earthly and bodily. Just as the popular sphere of the marketplace was opposed to the official sphere of the court and the church, so the grotesque body of the popular sphere opposes the monumental body of the official sphere. For Bakhtin (1984, 317–318):

> [T]he artistic logic of the grotesque image ignores the closed, smooth, and impenetrable surface of the body and retains only its excrescences (sprouts, buds) and orifices, only that which leads beyond the body's limited space or into the body's depths.

Similarly the artistic logic of the grotesque image ignores the closed, smooth, and impenetrable surface of the landscape and retains only its excrescences (outcrops, tors) and orifices, only that which leads beyond the limited space of the *polis*, or the public sphere of politics, and the *oikos*, or private sphere of the household, into the biosphere of the earth's depths, into the underworld and the depths of the 'jungle' (rainforest) and the 'swamp' (wetland).

7.1 Monstrous Alligators and Crocodiles

Thoreau was an avid reader of travel writing about exotic places that included nature writing about their exotic plants and animals. Thoreau read the work of William Bartram who travelled to a Florida swamp in the late eighteenth century where and when he encountered alligators. Alligators, for Bartram, are monsters who inhabit the grotesque lower earthly stratum of the swamp. Bartram encountered "the subtle, greedy alligator" about to devour "the voracious trout," the eater eaten, the preyer predated by a larger predator:

> His enormous body swells. His plaited tail brandished high floats upon the lake. The waters like a cataract descend from his opening jaws. Clouds of smoke issue from his dilated nostrils. The earth trembles with his thunder. (Bartram 1998, 75)

Two hundred years later Glasgow took this last sentence as the subtitle for his social history of the American alligator and suggests that Bartram "certainly wins, hands down, any contest for colourful alligator writing" (Glasgow 1991, 31). There is certainly plenty of that in the intervening two centuries as Glasgow's study shows. Bartram's description is a lurid traveller's tale complete with exaggeration that the vapour is smoke for which he has a biblical warrant (Job 41: 20), though in a later encounter he describes how "the vapour ascends from his nostrils like smoke" (Bartram 1998, 82). It would only take the addition of fire from the alligator's mouth and wings on its back to complete the cultural construction of the alligator as a dragon. Bartram's alligators are, as Slaughter (1996, 200) points out, "monsters on a heroic scale. Bartram's drawing, *The Alegator* [*sic*] of *St Johns* [now housed in the Natural History Museum in London], resembles medieval representations of the dragons battled by St George." Bartram on a couple of occasions refers to "the monster" (Bartram 1998, 77, 80). Bartram is a latter day St George to the alligator's dragon with some considerable biblical and cultural warrant for making this association (Giblett 2019, 18).

The swamp is a secular underworld into which the hero of the modern adventure romance has to descend and in which he has to overcome monsters (Giblett 1996). Bartram's travels involve him not only descending into Florida swamps but also overcoming their resident monsters, alligators—as he does later to emerge "victorious, or at least [having] made a safe retreat" (Bartram 1998, 80) as he concedes. Bartram leaves us in no doubt that we are in the underworld when he describes "the boiling surface of the lake" (75) created by two alligators engaged in "horrid combat" (76). The alligator swamp is a hellish, hot place and a place where the elements of water (lagoon) and fire (or heat) mix, as do the elements of air, earth, and water created by the alligator itself. The "dreadful roar" (76) or the "terrifying roar" of the alligator "resembles very heavy distant thunder, not only shaking the air and waters, but causing the earth to tremble" (82).

The alligator mixes the elements of earth, air, water, and fire (thunder and lightning are the fire in the sky) just as the wetland does more generally (see Giblett 2018, chapter 1; see also Giblett 1996). Instead of these elements staying put in their proper place, the alligator and the wetland mix them up and violate the order of things that assigns them to a fixed and stable category. Even the smoke/vapour exhaling from the alligator's nostrils upsets the distinction between air, water, and fire (where there is smoke, there's fire). The fact that the alligator in the water "resembles, at a distance, a great chunk of wood floating about" (Bartram 1998, 82) means that it pretends to be solid and earthy when in fact it is monstrous and slimy.

And orally sadistic to boot. If there is one feature of the alligator that more than any other arouses dread and horror for Bartram it is the fear of being "dragged out of the boat and instantly devoured" (Bartram 1998, 76) by the greedy monster. Its jaws are prodigious with the upper jaw opening at right angles to the lower one according to Bartram (82) but mistakenly for Glasgow (1991, 31). Bartram also found that they are capable of "belching floods of water" (Bartram 1998, 76). The jaws of the alligator emit terrifying roars and belch water, and so give out noise and liquid. They also take in liquid in the form of water and blood, and solid in the form of flesh, animal or human. Their jaws give and take. Their jaws are a transportational zone between the inside and outside of the body through which matter passes.

Alligators for Bartram are objects of horror who potentially could devour him as an abject of consumption. Bartram reproduces the hierarchies of norm society. John Muir followed in Bartram's swampy footsteps when he ventured into a Florida swamp nearly a century after Bartram. He did not follow in them by typecasting alligators as monsters who inhabit the grotesque lower earthly stratum of the swamp. Rather he portrayed alligators as God's creatures who have their place and role in nature. They were crucial for the development of his biocentric ethic. Muir's *A Thousand Mile Walk to the Gulf* was published posthumously in 1916 relating a journey he had undertaken almost 50 years earlier. In 1867 Muir, "one of the seminal figures in the history of modern environmental thought" according to Frederick Turner in his introduction to the Penguin Nature Library edition of this book, encountered an alligator "on the margin of a stagnant pool" in a Florida swamp which prompted Muir to reflect:

> These independent inhabitants of the sluggish waters of this low coast cannot be called the friends of man. [...] Many good people believe that alligators were created by the Devil, thus accounting for their all-consuming appetite and ugliness. But doubtless these creatures are happy and fill the place assigned them by the great Creator of us all. Fierce and cruel they appear to us, but beautiful in the eyes of God. They, also, are his children, for he hears their cries, cares for them tenderly, and provides their daily bread. The antipathies existing in the Lord's great animal family must be wisely planned, like balanced repulsion and attraction in the mineral kingdom. How narrow we selfish, conceited creatures are in our sympathies! How blind to the rights of all the rest of creation! With what dismal irreverence we speak of our fellow mortals! Though alligators, snakes, etc. naturally repel us, they are not mysterious evils. They dwell happily in these flowery wilds, are part of God's family, unfallen, undepraved, and cared for with the same species of tenderness and love as is bestowed on angels in heaven or saints on earth. I think that most of the antipathies which haunt and terrify us are morbid productions of ignorance and weakness. I have better thoughts of those alligators now that I have seen them at home. Honorable representatives of the great saurians of an older creation, may you long enjoy your lilies and rushes, and be blessed now and then with a mouthful of terror-stricken man by way of dainty! (Muir 1992, 98–99)

Here endeth the sermon on "how I learned to stopped hating alligators, and learnt to love them." Only a true greenie would hug an alligator metaphorically as readily as they would a tree in actuality. Perhaps needless to say, Muir was not a mouthful of terror-stricken humanity for the alligators of a Florida swamp. He may have thought differently and not have been quite so jaunty if he had been.

Alligators in a Florida swamp function in both Bartram's and Muir's texts as figures for the British colonial and American cultural unconscious. Florida and the American south more generally with its swamps have been repressed in the collective psyche of the American north and have functioned as a figure of the primitive and backward (see Giblett 1996 and references). Relatedly, American wetlands have been drained and their creatures commodified from early colonial times. In Bartram's case he is returned to this repressed only to try to continue to repress it; in Muir's, he embraces it—albeit metaphorically. Muir also embraced the life of a

Canadian marsh (as will see in a later chapter). He also embraced the hierarchies of aesthetics in national parks and wilderness (as will also see in a later chapter). In Thoreau's case, he embraced the repressed of the grotesque lower earthly stratum of the swamp (as will also see in a later chapter) and subverted the aesthetic hierarchy of the upper earthly stratum of the city over the swamp (as we will see in the present chapter). He also repressed the grotesque lower bodily stratum (as we will also see in the present chapter).

Thoreau upset the conventional view of the dominant cultural paradigm at the time that swamps were poisonous by parodying it in his reference to the "rank and venomous luxuriance in this swamp" (Thoreau 1962, *IX*, 60). Even at the worst of times he could prescribe a swamp cure, such as when he wrote in his journal that "[w]hen life looks sandy and barren, is reduced to its lowest terms, we have no appetite, and it has no flavor, then let me visit such a swamp as this, deep and impenetrable, where the earth quakes for a rod around you at every step, with its open water where swallows skim and twitter" (Thoreau 1962, *IV*, 231). When desire is diminished and life is dissatisfying, both figured here orally, the quaking zone of the swamp has depths and softness that the shallowness of its waters belie, and that the depths of the lake cannot dream of. Thoreau values precisely those usual denigratory connotations in the dominant cultural paradigm that attach to the "depth" or horizontal extension and impenetrability of the swamp.

Thoreau also upsets the usual dissociation between swamp and city by seeing the city as swamp and so subverts the unfavourable connotations of the swamp as a place of darkness, disease, and even death. Rather than the swamp, Thoreau saw "society," "civilization," and the modern city as bearers of disease, or perhaps more precisely he saw the modern city as swamp in the conventional sense of an uncanny and unhomely place of disease and horror, and saw simultaneously the swamp as canny and homely, as postmodern dwelling in the unconventional sense of a homely, but also wild (homely *because* wild) place. Thoreau (1962, *II*, 47) wrote in his journal that he could "see less difference between a city and some dismallest swamp then formerly. It is a swamp too dismal and dreary, however, for me." The city for Thoreau is dismal, not the city. He upsets and overturns the conventional association between swamp and dismalness, and the city not being dismal. Although he would prefer the swamp as swamp over the city as swamp, he nevertheless goes on to make a finer distinction: "I would prefer even a more cultivated place, free from miasma and crocodiles." And wouldn't we all? He still prefers the swamp to the city, and the native quaking zone of the swamp where crocodiles and alligators live to the feral quaking zone of the city with its swampy slums (Giblett 2009, 1–54). The alligator for Thoreau (1982, 543) "comes out of the mud with quakings of the earth."

Perhaps Thoreau was wise to prefer "a more cultivated place" free of crocodiles, though this did not mean that he preferred the city to the swamp, especially as the former could be as dangerous in common parlance as the latter. The city as swamp for Thoreau (1982, 350) in *Walden* had its own diseases and horrors:

> [L]et us settle ourselves, and work and wedge our feet downward through the mud and slush of opinion, and prejudice, and tradition, and delusion, and appearance, that alluvion which covers the globe, through Paris and London, through New York and Boston and Concord, through church and state, through poetry and philosophy and religion, till we come to a hard bottom and rocks in place which we call *reality*. (His emphasis)

The four cities and the town mentioned were founded in, or next to, and expanded into, marshes or swamps.[1]

Thoreau is here using "the mud" as a pejorative figure for noise. Mud is also the home of eels and even 'in-forms' them. Writing about his river journey, Thoreau (1998, 28) describes in *A Week on the Concord and Merrimack Rivers:* "[T]he Common Eel […] *Muraena Bostoniensis* [American Eel (*Anguilla rostrata*)], the only species of eel known in the State [of Massachusetts as], a slimy, squirming creature, informed of mud." Eighteen years after his river journey Thoreau (1962, *X*, 114) came across a dead, two-foot long eel hanging on a bough and exclaimed in his journal "what a repulsive and gluttonous-looking creature, with its vomer made to plow the mud and wallow in filth. […] It is more repulsive to me than a snake, and I think must be less edible."

Writing in his journal the year before *Walden* was published "the mud" is "Nature's womb" (Thoreau 1962, *V*, 499). "The mud" is an ambivalent substance, a double figure. So, let us settle ourselves, and work and wedge our feet downward through the noise of the muddy media (including social media) and through the cities built in muddy swamps, and so

[1] For discussion of Paris, London, New York, and Boston, see Giblett (2016, chapters 3, 4, 9, and 10).

on, till we come to a soft bottom and mud in place which we call *reality*. Carnivalesque subversion in and with the grotesque body and places, such as swamps, involves “Nature’s womb” of mud giving birth to new life.

7.2 Grotesque Swamps

Swamps are grotesque places that subvert the conventions and norms of aesthetics, such as the sublime, the beautiful, and the picturesque (Bakhtin 1984, 35; Giblett 1996, chapter 2). Bakhtin (1984, 43) argues that the sublime and grotesque “complete each other.” One is not possible without the other. They are complementary, mutually constitutive, and in incessant struggle. In 1827 Victor Hugo reflected at length in the “Preface” to his play *Cromwell* on the grotesque in relation to the sublime and the beautiful. For Hugo (1910, 345) “the modern muse […] will realize that everything in creation is not humanly *beautiful*, that the ugly exists beside the beautiful, the unshapely beside the graceful, the grotesque on the reverse of the sublime, evil with good, darkness with light.” For Hugo (1910, 345–346) the modern muse in realizing this is merely “doing as nature does, mingling in its creations—but without confounding them—darkness and light, the grotesque and the sublime; in other words, the body and the soul, the beast and the intellect.”

Hugo (1910, 346) concludes later in his “Preface” that “it is from the fruitful union of the grotesque with the sublime that the modern spirit is born.” The modern city, including Paris, particularly the modern swamp city and marsh metropolis, is, however, not born from the same fruitful union as the novel of the grotesque with the sublime, the monstrous with the monumental, the abjective with the objective and subjective, but is born of the fatal union of city and swamp, marsh and metropolis, in which the mother swamp dies, or is killed, in giving birth to the city, in Hausmann’s case by caesarean section or still-birth by abortion. The truly postmodern city would achieve a rapprochement and reconciliation between the modern city and the pre-modern mother goddess of the swamp.

Hugo’s manifesto on the grotesque and the sublime did not go unremarked by Bakhtin who relates how, “in the 1820s, there was a revival of grotesque imagery in French Romanticism. We find an interesting presentation of the problem, and one typical of the French Romantic spirit, by Victor Hugo, first in his preface to *Cromwell*” (Bakhtin 1984, 43). Bakhtin (1984, 43) goes on to note that for Hugo “the aesthetics of the grotesque are to a certain extent the aesthetics of the monstrous." Bakhtin (1984, 43) then critiques Hugo's "Preface": "but at the same time Hugo reduces the intrinsic value of the grotesque by declaring that it is a means of contrasting the sublime. The two complete each other" for Hugo in holistic and functionalist complementarity, whereas for Bakhtin the grotesque counters and subverts the sublime in unceasing socialist struggle both in the body and the city, and in the city as body (see Giblett 2008, chapters 4 and 5).

Thoreau came to terms with grotesque places, such as swamps, yet struggled to come to terms with the grotesque body, his own and others', such as eels, and with the grotesque stratum of his own mind. The post-Romanticist Thoreau preferred swamps to lawns, the quaking zone of swamps in the country to the manicured surface of lawns in the town and city. When he waded through the mud and water of a swamp or marsh he was embodied in the watery body of the earth. By contrast, when the Romanticist and Transcendentalist Thoreau of *Walden* looked at the clouds reflected in water he was sublimated from solid earth through liquid water into gaseous air. For Thoreau (1982, 438) "in such transparent and seemingly bottomless water, reflecting the clouds, I seemed to be floating through the air as in a balloon." Looking at the clouds reflected in water is sublimated transcendence, whereas for Thoreau in writing in *Walden* lying on his back in his boat floating on the pond is "dreaming awake":

> I have spent many an hour, when I was younger, floating over its [Walden Pond's] surface as the zephyr willed, having paddled my boat to the middle, and lying on my back across the seats, in a summer forenoon, dreaming awake, until I was aroused by the boat touching the sand, and I arose to see what shore my fates had impelled me to; days when idleness was the most attractive and productive industry. Many a forenoon have I stolen away, preferring to spend thus the most valued part of the day; for I was rich, if not in money, in sunny hours and summer days, and spent them lavishly; nor do I regret that I did not waste more of them in the workshop or the teacher's desk. (Thoreau 1982, 440)

Dreaming awake is a meditative state of calmness in which the dreamer is woke to the temporal work-discipline of industrial capitalism (to cite Thompson 1968). Contra Benjamin Franklin for whom "time is money" with its implicit financial injunction to spend time working for money and its moral imperative that time wasted is money lost, richness for Thoreau is time spent "dreaming awake" and whiling away the time in his supine posture in this place of Walden Pond not working for money in the most valued part of the day.

By contrast with looking into a lake and measuring the depth of one's own nature, when Thoreau is dreaming asleep this depth can come to the surface, such as when he wrote in *A Week*:

> Dreams are the touchstones of our characters. We are scarcely less afflicted when we remember some unworthiness in our conduct in a dream, than if it had been actual, and the intensity of our grief, which is our atonement, measures the degree by which this is separated from an actual unworthiness. For in dreams we but act a part which must have been learned and rehearsed in our waking hours, and no doubt could discover some waking consent thereto. If this meanness had not its foundation in us, why are we grieved at it? In dreams we see ourselves naked and acting out our real characters, even more clearly than we see others awake. But an unwavering and commanding virtue would compel even its most fantastic and faintest dreams to respect its ever-wakeful authority; as we are accustomed to say carelessly, we should never have *dreamed* of such a thing. Our truest life is when we are in dreams awake. (Thoreau 1998, 239–240; his emphasis)

Dreaming asleep for Thoreau manifests the depths of the grotesque lower bodily and mental strata, and the monstrous self, certainly not his truest self. His disembodied self looking at the clouds reflected in a lake, transcending the earth, and being sublimated into the upper mental stratum of the sublime was not his truest self either. He found his embodied self entering the grotesque lower earthly stratum of the sacral swamp and being desublimated into the slimy body of the earth and his own body. He found himself, the depths of his own nature, and nature looking into a pond and a lake, and his "truest self" "dreaming awake" lying on his back in his boat floating on Walden Pond. These selves mediate the upper and lower strata and achieve rapprochement between them. "Dreaming awake and asleep" (Thoreau 1962, *XI*, 287) mediates and manifests the transitions of the self between the sublime upper mental stratum, the grotesque lower bodily and mental strata of one's own nature, and the grotesque lower earthly strata of nature, especially the swamp in psychopolitical wetland ecology.

The swamp for Thoreau (1962, *IX*, 394) writing much later in his journal is shallow and soft, the first birth of nature:

> [T]hat central meadow and pool in Gowing's Swamp is its very navel, *omphalos*, where the umbilical cord was cut that bound it to creation's womb. Methinks every swamp tends to have or suggests such an interior tender spot. The sphagnous crust that surrounds the pool is pliant and quaking, like the skin or muscles of the abdomen; you seem to be slumping into the very bowels of the swamp.[2]

In other words, slumping into the abject body of the earth. The swamp as creation's womb is the central tenet of Thoreau's evolutionary biology and his conservation counter-creationist theology of swampy evolution. Gowing's Swamp dries up and gets wet during the seasonal cycle of the rising and falling waters of the underlying aquifer, which is typical of wetlands. Thoreau (1958a, 165, 173) wrote in his journal that "the springs of life flow in ceaseless tide below" in the "lower strata of nature," the grotesque lower stratum of the earth.

Another way in which Thoreau upset the standard rhetoric of swamp-speak in which the swamp is a place of death and disease is that for him the swamp is the stuff of life and death. Indeed, for Thoreau writing in a letter to Emerson "death is only the phenomenon of the individual or class. Nature does not recognize it" (Thoreau 1958b, 64; 2013, 104). Moreover, for him in the same letter death is "a law and not an accident—It is as common as life [... T]he law of their [flowers] death is the law of new life" (1958b, 64; 2013, 105). The swamp, as with nature generally, upsets the hard and fast distinction between life and death. Thoreau (1962, *XIV*, 109; 1993, 100–101) inverts the morbid Christian orthodoxy of the line from the Anglican/Church of England *Book of Common Prayer* that "in the midst of life, we are in death" by maintaining how "in the midst of death, we are in life." In the middle of death in the swamp, we are in life. The swamp as marrow is constantly being renewed by the life-blood of the earth and constantly renews the bones of the body of the earth that give it structure. This is Thoreau's evolutionary biology and his conservation counter-theology of swamps rolled into a nutshell.

[2] Carl Bode (1982, 686) coyly or squeamishly excludes the first sentence and last phrase when he quotes this passage from Thoreau's journal. Thoreau drew a rough illustration of the quaking ground and sphagnum of Gowing's Swamp earlier in his journal; see Thoreau (1962, *VI*, 467).

7.3 Monstrous Quaking Zone

Thoreau was a frequent visitor to Gowing's Swamp. It is within a mile (about a kilometre) of Thoreau's birthplace (see map of Concord, Mass., and Index, Thoreau 1962, *XIV*). Of the hundreds of places and sites he wrote about, Gowing's Swamp is the closest to his birthplace. He wrote about it repeatedly and drew two maps of it in his journal where he described "its quaking ground" and "*navel* pool in the centre" (1962, *VI*, 467; *X*, 198 and *XIII*, 125 his emphasis). Swamps are maternal places where new life is born. They are places of the monstrous-maternal.

The swamp is also a native quaking zone of horror as Thoreau (1962, *VI*, 467) also suggested in his journal when he examined the middle of Gowing's Swamp and found an open pool of water nearly full of sphagnum and other plants, but he could not see them "on account of the danger of standing on the quaking ground [...] [and] of quaking sphagnum in which I sink eighteen inches in water [...] where I fear to break through." The quaking zone is a dangerous place that elicits fear and trembling because it not only mixes land and water creating mud and slime but also threatens to engulf the whole body fully in its slimy depths, immerse the subject in the abject, and mark the surface of the body with mud.

Thoreau (1962, *VI*, 479) found these aspects again five days later while "picking our way over quaking meadows and swamps and occasionally slipping into the muddy batter midleg deep." The native quaking zone does not give a secure and stable place for standing upright, having a fixed point of view, and observing the scene before one—preferably a pleasing prospect stretching before one as a subject with an object. The native quaking zone threatens to engulf the body in the slimy depths of the abject and return the subject to the watery womb from which it was born. Outside this zone, however, the view from the edge is another matter. Thoreau (1962 *XIV*, 301–302) wrote in his journal about "the open swamp [...] where the surface quakes for a rod around" that "there is no wilder and richer sight than is afforded from such a point of view of the edge of a [...] swamp." The native quaking zone affords visual pleasure from a distance, but provokes tactile horror up close and personal.

The monstrous lives and lurks in the grotesque lower strata of the earth, the body, and the mind where life and death are intertwined, as they were for Thoreau. Monstrosity for Georges Canguilhem (2005, 188), "not death, is the counter-value to life." The monster is not valueless nor dead, but rather, as for Canguilhem (2005, 188), it is "the living being of negative value" and this value is that of a foil. The monstrous plays a valuable cultural role and serves a valuable cultural function. Monsters for Michel Foucault (1970, 155) form "the background noise, as it were, the endless murmur of nature." The monstrous is black noise in the channels of communication that make it possible and threaten it at same time.[3] The monstrous threatens not only the stability of the symbolic order but also its child and progenitor the semiotic subject—the viewer, the reader, the maker of meaning. For Barbara Creed (1986, 64) the viewing subject confronted by what she calls "the sight of the monstrous" is "put into crisis—boundaries, designed to keep the abject at bay, threaten to disintegrate, collapse." The dyke against the flood, the bulwark against the tide, the lock against the canal all threaten to crumble and collapse and swamp the subject with the monstrous-maternal and the slimy depths of the body and the earth of nurturing nature.

References

Bakhtin, M.M. 1984. *Rabelais and His World*. Translated by H. Iswolsky. Bloomington, IN: Indiana University Press. First published in Russian in 1965.

Bartram, William. 1998. In *Travels: Naturalist's Edition*, ed. Francis Harper. Athens, GA: University of Georgia Press. First published 1791.

Bode, Carl. 1982. Epilogue by the Editor. In *The Portable Thoreau*, ed. Carl Bode, 683–696. New York: Viking Penguin.

Canguilhem, Georges. 2005. Monstrosity and the Monstrous. In *The Body: A Reader*, ed. Mariam Fraser and Monica Greco, 187–193. London: Routledge.

Creed, Barbara. 1986. Horror and the Monstrous-Feminine: An Imaginary Abjection. *Screen 27* (1): 44–70.

Foucault, Michel. 1970. *The Order of Things: An Archaeology of the Human Sciences*. London: Tavistock.

Giblett, Rod. 1996. *Postmodern Wetlands: Culture History Ecology*. Edinburgh: Edinburgh University Press.

Giblett, Rod. 2008. *Sublime Communication Technologies*. Basingstoke: Palgrave Macmillan.

Giblett, Rod. 2009. *Landscapes of Culture and Nature*. Basingstoke: Palgrave Macmillan.

Giblett, Rod. 2016. *Cities and Wetlands: The Return of the Repressed in Nature and Culture*. London: Bloomsbury.

[3] For further discussion of black noise (and white noise), see Giblett (2008, chapter 1).

Giblett, Rod. 2018. *Environmental Humanities and Theology: Ecoculture, Literature and the Bible*. London: Routledge.

Giblett, Rod. 2019. *Environmental Humanities and the Uncanny: Ecoculture, Literature and Religion*. London: Routledge.

Glasgow, Vaughn L. 1991. *A Social History of the American Alligator: The Earth Trembles with his Thunder*. New York: St Martin's Press.

Hugo, Victor. 1910. Preface to *Cromwell*. In *Prefaces and Prologues to Famous Books: With Introductions and Notes, the Harvard Classics, Volume 39*, ed. C. W. Eliot, 337–387. New York: P. F. Collier & Son. First published in French in 1827.

Morson, Gary Saul, and Caryl Emerson. 1990. *Mikhail Bakhtin: Creation of a Prosaics*. Stanford: Stanford University Press.

Muir, John. 1992. *A Thousand-Mile Walk to the Gulf*. New York: Penguin. First published 1916.

Slaughter, Thomas. 1996. *The Natures of John and William Bartram*. New York: Random House.

Thompson, E. P. 1968. *The Making of the English Working Class*. Harmondsworth: Penguin.

Thoreau, Henry David. 1958a. *Consciousness in Concord: The Text of Thoreau's Hitherto 'Lost Journal' (1840–41), Together with Notes and a Commentary by Perry Miller*. Boston: Houghton Mifflin.

Thoreau, Henry David. 1958b. In *The Correspondence*, ed. Walter Harding and Carl Bode. New York: New York University Press.

Thoreau, Henry David. 1962. In *The Journal 1837–1861, Volumes I–XIV*, ed. Bradford Torrey and Francis Allen. New York: Dover.

Thoreau, Henry David. 1982. In *The Portable Thoreau*, ed. Carl Bode. New York: Viking Penguin.

Thoreau, Henry David. 1993. In *Faith in a Seed*, ed. Bradley Dean. Washington, D.C.: Island Press/Covelo, California: Shearwater Books.

Thoreau, Henry David. 1998. In *A Week on the Concord and Merrimack Rivers*, ed. H. Daniel Peck. New York: Penguin. First published in 1849.

Thoreau, Henry David. 2013. In *The Correspondence: Volume 1: 1834–1848*, ed. Robert Hudspeth. Princeton, NJ: Princeton University Press.

White, Allon. 1993. Prosthetic Gods in Atrocious Places: Gilles Deleuze/Francis Bacon. In *Carnival, Hysteria and Writing: Collected Essays and Autobiography*, 160–177. Oxford: Clarendon Press.

CHAPTER 8

Nature Writing Is Factual and 'Non-fiction' (Faction)

Nature writing is usually classified in that curious category of 'non-fiction.' It is categorized as what it is not, as an absence or lack, rather than what it is, a presence or being. Non-fiction is "the leftover term apotheosizing fiction," as Solnit (2014, 4) calls it. Non-fiction is what is left over after the consumption of the divine dish of fiction in the carve up of the feast of books and writing. Non-fiction is the other of fiction, without which it could not exist. Fiction requires non-fiction to constitute itself. Without 'non-fiction,' fiction could not and would not exist. Nature writing does not require fiction to exist. It is heterogeneous, and mixed with other genres (as we saw in Chap. 8). Nature is a matter of fact, not fiction, yet writing about it is often both factual and fictional, and nature has its own stories to tell that are sometimes fictional. The line between fact and fiction is often blurred, and deliberately so in nature and nature writing. Nature writing is faction.

The most apotheosized genre of fiction is the novel. It has been valorized by novelists, critics, and publishers over every other genre, or marketing category. Novels share some characteristics with nature writing, such as both are a long prose narrative with a beginning, middle, and an end made up of shorter episodes with an overarching plot or trajectory. They also differ. Novels are driven by fictional characters and invented scenarios that can be realistic representations or thinly disguised versions of both taken from ordinary, everyday life. Novels also usually occur within

R. Giblett, *The Politics of Nature and Environmental Writing*, Transdisciplinary Environmental Humanities 1,
https://doi.org/10.1007/978-3-032-04427-3_8

the narrow and claustrophobic confines of a domestic, indoor setting with an outdoor landscape as a background for context, not a player or actor in its own right. Nature writing is about the factual characters of the writer and other beings living in real places. Nature writing also takes place in outdoor settings with human and non-human agents and actors in the ordinary and extraordinary life of nature.

8.1 Nature Writing and Story

For instance, Thoreau wrote a story in *Walden* about:

> an artist in the city of Kouroo who was disposed to strive after perfection. One day it came into his mind to make a staff. Having considered that in an imperfect work time is an ingredient, but into a perfect work time does not enter, he said to himself, It shall be perfect in all respects, though I should do nothing else in my life. [...] As he made no compromise with Time, Time kept out of his way, and only sighed at a distance because he could not overcome him. (Thoreau 1982, 565)

In his search for the perfect material for his stick, his friends deserted him. Before he found the suitable material, the city of Kouroo was a ruin. And so on:

> [B]efore he had given it the proper shape the dynasty of the Candahars was at an end. By the time he had smoothed and polished the staff Kalpa was no longer the pole-star;
>
> and ere he had put on the ferule and the head adorned with precious stones, Brahma had awoke and slumbered many times. (Thoreau 1982, 565–566)

Thoreau then asks:

> But why do I stay to mention these things? When the finishing stroke was put to his work, it suddenly expanded before the eyes of the astonished artist into the fairest of all the creations of Brahma. He had made a new system in making a staff, a world with full and fair proportions; in which, though the old cities and dynasties had passed away, fairer and more glorious ones had taken their places. And now he saw by the heap of shavings still fresh at his feet, that, for him and his work, the former lapse of time had been an illusion, and that no more time had elapsed than is required for a single

> scintillation from the brain of Brahma to fall on and inflame the tinder of a mortal brain. The material was pure, and his art was pure; how could the result be other than wonderful? (Thoreau 1982, 566)

The source of this story, if indeed it has a source outside of Thoreau's creative imagination, has perplexed Thoreauvian scholars, as Walter Harding notes in his annotated edition of *Walden* (Thoreau 1995, 317 n3). Perhaps it was designed to send his readers on a wild goose chase on which some of them have gone, never to come back. Thoreau gives the story an air of authority, authenticity, and antiquity that it might otherwise lack. He blurs the line between fiction and 'non-fiction.' The prosaic facticity of chronological and terrestrial time becomes a poetic fiction of timeless and cosmic eternity.

This story is similar to Thoreau's cock and bull story of his search for "a hound, a bay horse, and a turtle dove" that he lost long ago and on whose trail he was still on:

> [M]any are the travellers I have spoken concerning them, describing their tracks and what calls they answered to. I have met one or two who had heard the hound, and the tramp of the horse, and even seen the dove disappear behind a cloud, and they seemed as anxious to recover them as if they had lost them themselves. (Thoreau 1982, 272)

This story has perplexed Thoreauvian scholars too, even more so than the story of the Kouroo artist, as Walter Harding notes in the "Appendix" devoted to it in his annotated edition of *Walden* (Thoreau 1995, 327–329). Harding discusses various explanations and gets Thoreau's joke when he concludes that "by this time it should be obvious that one needs a sense of humor in considering all these critical reactions," including making up the story on the spot as an excuse for trespassing (Thoreau 1995, 329). The whole of *Walden* can be read as making up an excuse for crossing the bounds of sense and genre, as well as trespassing on the time and attention of the reader. He announced in the very beginning of *Walden* that his reason for being about to "obtrude my affairs so much on the notice of my readers" in writing the book was to answer the obtrusive and prying inquiries of "his townsmen" (Thoreau 1982, 258). Thoreau and *Walden* were obtrusive too and gave back as good as he got.

Harding concludes finally "—and to be serious once again—there is no unanimity on interpretation of these symbols, and each reader is free to interpret them as he wishes." Harding is following Thoreau who gave some directions in the conclusion to *Walden* for how it should be read, including that the reader should entertain and perform different interpretations of it. He quotes a French history of Hindu literature[1] that discusses the mystic and poet Kabir: "'the verses of Kabir have four different senses; illusion, spirit, intellect, and the exoteric doctrine of the Vedas'; but in this part of the world it is considered a ground for complaint if a man's writings admit of more than one interpretation" (Thoreau 1982, 564). The same complaint still persists in most parts of the world.

The scholars have not been able to track down and find a source for the stories of the Kouroo artist and "a hound, a bay horse, and a turtle dove." They have been more successful with Kabir and Krishna. Thoreau quotes Damodara in *Walden* who said "there are none happy in the world but beings who enjoy freely a vast horizon." Damodara is another name for Krishna. Harding notes in his annotated edition of *Walden* that Thoreau is quoting and probably translating his own French translation of *Harivansa*, now transliterated as *Harivamsa*, an important work of Sanskrit literature (Thoreau 1995, 84, n5; see also 82, n4). Thoreau did indeed translate the French translation into English, a limited edition of which was edited and introduced by Arthur Christy and published in 1932 that escaped Harding's notice (Thoreau 1932). This edition includes a facsimile of Thoreau's handwritten manuscript now housed in the Houghton Library at Harvard University (Jeffrey Cramer, email to author August 30, 2023). Thoreau's translation is now included in the Princeton University Press volume of Thoreau's translations in its standard edition of his *Writings*. The facts of scholarship and the fancies of fiction are both present in *Walden*. Nature writing is faction.

The prosaic facts of Thoreau's everyday life are also presented in *Walden*, such as in his accountancy:

> The exact cost of my house, paying the usual price for such materials as I used, but not counting the work, all of which was done by myself [except for his 'acquaintances' helping to raise the frame], was as follows; and I give the details because very few are able to tell exactly what their houses cost, and fewer still, if any, the separate cost of the various materials which compose them:—

[1] As noted by Walter Harding (Thoreau 1995, 316 n2).

> Boards .... $ 8.03+, mostly shanty boards.
> Refuse shingles for roof sides ... 4.00
> Laths 1.25
> Two second-hand windows
> with glass 2.43
> One thousand old brick 4.00
> Two casks of lime .... 2.40 That was high.
> Hair .... 0.31 More than I needed.
> Mantle-tree iron 0.15
> Nails 3.90
> Hinges and screws 0.14
> Latch 0.10
> Chalk 0.01
> Transportation .... 1.40} I carried a good part on my back.
> In all .... $28.12+
> These are all the materials, excepting the timber, stones, and sand, which I claimed by squatter's right. (Thoreau 1982, 303)

Thoreau gave a similar accounting of his outgoings and income from cultivating, harvesting, and selling his beans (Thoreau 1982, 412).

8.2 The Facts of Life and Nature

Thoreau was also concerned with the facts of life and nature in his first book, *A Week on the Concord and Merrimack Rivers*, about the river journey he and his brother took over two weeks in a boat he and his brother built in a week. More than just a boys-own adventure story of messing about in a boat, it is a meditation on nature, including rivers and meadows. Following the course of Thoreau's riverine journey, on August 31, 2003, John McPhee and his companion began to re-trace by canoe Thoreau's and his brother's journey by boat along the same rivers beginning on August 31, 1839. Thoreau's and McPhee's journeys began in the town of Concord, Massachusetts, and then traversed the wetland of "the Great Meadows" which "have levelled a fertile and juicy place in nature" as Thoreau puts it in *A Week* (Thoreau 1998, 7) and which are described by McPhee (2004, xiv) as "part floodplain, part swamp."

Meadows for Thoreau are also quaking zones, such as when he refers to "the quaking meadow ground" (Thoreau 1998, 32). These are places where the earth trembles, and the body does too (as we will see in the following chapter of this book). McPhee is a great writer of 'non-fictional' or factual 'novels' of true stories about American land, landscapes, and their peoples. Following in McPhee's and Thoreau's footsteps, or in their wake, *Middlemarsh* is a factual novel of true stories about Australian land, landscapes, and their peoples in mid-western Victoria (Giblett 2023). *Middlemarsh* is structured around the course of the Hopkins River with a beginning, middle, and an end. It is also structured into seven chapters around the days of the week, like Thoreau's *A Week on the Concord and Merrimack Rivers*.

Thoreau's and McPhee's journeys by boat and canoe were up and down the Concord and Merrimack Rivers. Thoreau (1998, 13) observed in *A Week* that rivers are "the natural highways of all nations." Two centuries before Thoreau, Blaise Pascal said "a river is a road that moves" (cited in Giblett 2023, 45). A river is an active road whose downstream flow can move vessels along its course providing some motive power for their propulsion. Rivers, as Christof Mauch and Thomas Zeller (cited in Giblett 2023, 45) put it, are "themselves agents, providers of energy [not only of, for example, hydroelectric power generation, but also in their own movement *à la* Pascal], and resources, and a driving force in history." Rivers are not passive roads on which vehicles move, but active roads whose flow downstream also moves resources, such as water and nutrients, as Thoreau observed and noted later.

More than a decade after his weeklong riverine journey, Thoreau (1962, *V*, 404) wrote in his journal of the Concord River that "[n]ature made a highway [...] through this town [...] bordered by the most fertile soil in the town, a tract most abounding in vegetable and in animal life." Rivers convey river floaters "on their bosom [...] through the most interesting scenery," as Thoreau (1998, 13) puts it in *A Week*. They also convey nutrients in the milky, lifegiving waters of their bosom. Along their banks and flanks and in their waters is "where the animal and vegetable kingdoms attain their greatest perfection," according to Thoreau (1998, 13) in *A Week*.

Rivers are also roads that afford different points of view and scenery for the river-borne traveller from their landbound counterparts. Thoreau (1998, 190) observed in *A Week* that "the river steals into the scenery it traverses without intrusion, silently creating and adorning it." The landbound road intrudes noisily, both in the construction and in usage phases, into the scenery it traverses, often creating a place and point of view to take in the scenery and make the land into landscape. This certainly was the case with the construction of roads in National Parks in the US and

Australia. The curving sweep of roads with stopping points to take in the spectacular view were created to entice the car-driving tourist to stop and experience the land as landscape, as dead, aesthetic surface and not as living, dynamic depths.

For Thoreau (1998, 12) at the beginning of his journey on the Concord and Merrimack Rivers "the sluggish artery of the Concord meadows steals [...] unobserved through the town, without a murmur or a pulse-beat." The meadows are about fifty miles long and convey "a huge volume of matter" along its natural highway. Meadows without a pulse are water in a meditative state. When Taoist Tai Chi Master Moy Lin-Shin meditated, his circulations became sluggish and his students could not find a pulse. Thoreau (1958, 176) wrote in his journal that "it is the slowest pulsation which is the most *vital*." By contrast with unobserved, sluggish meadows, rivers typically pulse with observable flowing matter. They are part of what Thoreau (1958, 209; 1982, 34) calls in his journal and in his review, *Natural History of Massachusetts*, "the great pulse of nature" which "vibrates by [or 'clearly'] and through each instant."

Thoreau does not distinguish greatly between the Concord meadows and Concord River. The river is a part of the meadows, and vice versa. Thoreau (1962 *XII*, 263) in one case wrote in his journal late in life that:

> you can [...] feel the pulse of our river only in the shallowest places, where it preserves some slight passage between the weeds. It faints and gives up the ghost in deeper places on the least adverse wind, and you would presume it dead a thousand times if you did not apply the nicest tests, such as a feather to the nostril of a drowned man. It is a mere string of lakes which have not made up their minds to be rivers.

The meadows supply the string of lakes with lifegiving water and nutrients. They are all just one continuous body of water in contiguous watery bodies. Water spreads across the ground and flows to its lowest point. These are facts of life and nature. Where a river ends in time and space, and where a billabong, meadow, or floodplain begins (as we saw with Bill Neidjie in Chap. 3), highlights the problematic nature of the terms 'floodplain' and 'river' applied to nature and water. This is a matter of cultural critique and the politics of water. They are bodies of water.

With or without a murmur or a pulse, all meadows and rivers, lagoons, and lakes are living waters, the lifeblood of the body of the earth. More than twenty years after his journey on the Concord and Merrimack Rivers,

Thoreau wrote in his journal that water is "the most living part of nature. This is the blood of the earth, and we see its blue arteries pulsing with new life now" (Thoreau 1962, *XIII,* 163). Conversely, the blood of the body pulses through veins and arteries like "those waters, which are carried by brooks and rivers over all the earth," as Thoreau (2001, 78) cites Sir Walter Raleigh to this effect. Blood is the lifegiving water of the body. Even in winter, water is still an artery of lifeblood. Thoreau (1962, *IX,* 224) wrote in his journal three years earlier about how a brook in winter was "open in the meadow there, an artery of black water in the midst of the snow." Thoreau (1962, *XIII,* 138) wrote three years later that the "black artery" of the river in winter is "the wrists, temples of the earth, where I feel its pulse with my eye. The living waters, not the dead earth. It is as if the dormant earth opened its dark and liquid eye upon us." Living black waters of a native wetland are not the dead black waters of a feral wet wasteland (or waste wetland).

8.3 The Facts and Figures of Life and Nature

The facts of non-human nature and the facts and fictions of human culture also coalesce in Thoreau's nature writing in his use of tropes and figures, such as in 'the sand foliage' episode in *Walden.* Industrial technology intruded into Walden Woods when the construction of a cut on the track for the Fitchburg Railroad produced a sand bank. Thoreau complained in *Walden* that the Fitchburg Railroad "touches the pond" and "the railroad has infringed on its border" (Thoreau 1982, 367, 441). The railroad is an invader that has soiled the hem of the pond and the fringing woods of its garment. Yet in doing so it gave Thoreau the opportunity to give a lengthy, and now famous, description of flowing sand foliage in *Walden* and connect it creatively and poetically to the morphology in common of plants and the human body:

> Few phenomena gave me more delight than to observe the forms which thawing sand and clay assume in flowing down the sides of a deep cut on the railroad through which I passed on my way to the village, a phenomenon not very common on so large a scale, though the number of freshly exposed banks of the right material must have been greatly multiplied since railroads were invented. The material was sand of every degree of fineness and of various rich colors, commonly mixed with a little clay. When the frost comes out in the spring, and even in a thawing day in the winter, the sand begins to

flow down the slopes like lava, sometimes bursting out through the snow and overflowing it where no sand was to be seen before. Innumerable little streams overlap and interlace one with another, exhibiting a sort of hybrid product, which obeys half way the law of currents, and half way that of vegetation. As it flows it takes the forms of sappy leaves or vines, making heaps of pulpy sprays a foot or more in depth, and resembling, as you look down on them, the laciniated, lobed, and imbricated thalluses of some lichens; or you are reminded of coral, of leopard's paws or birds' feet, of brains or lungs or bowels, and excrements of all kinds. It is a truly *grotesque* vegetation, whose forms and color we see imitated in bronze, a sort of architectural foliage more ancient and typical than acanthus, chiccory, ivy, vine, or any vegetable leaves; destined perhaps, under some circumstances, to become a puzzle to future geologists. The whole cut impressed me as if it were a cave with its stalactites laid open to the light. The various shades of the sand are singularly rich and agreeable, embracing the different iron colors, brown, gray, yellowish, and reddish. When the flowing mass reaches the drain at the foot of the bank it spreads out flatter into *strands*, the separate streams losing their semi-cylindrical form and gradually becoming more flat and broad, running together as they are more moist, till they form an almost flat *sand*, still variously and beautifully shaded, but in which you can trace the original forms of vegetation; till at length, in the water itself, they are converted into banks, like those formed off the mouths of rivers, and the forms of vegetation are lost in the ripple marks on the bottom.

The whole bank, which is from twenty to forty feet high, is sometimes overlaid with a mass of this kind of foliage, or sandy rupture, for a quarter of a mile on one or both sides, the produce of one spring day. What makes this sand foliage remarkable is its springing into existence thus suddenly. When I see on the one side the inert bank—for the sun acts on one side first—and on the other this luxuriant foliage, the creation of an hour, I am affected as if in a peculiar sense I stood in the laboratory of the Artist who made the world and me—had come to where he was still at work, sporting on this bank, and with excess of energy strewing his fresh designs about. I feel as if I were nearer to the vitals of the globe, for this sandy overflow is something such a foliaceous mass as the vitals of the animal body. You find thus in the very sands an anticipation of the vegetable leaf. No wonder that the earth expresses itself outwardly in leaves, it so labors with the idea inwardly. The atoms have already learned this law, and are pregnant by it. The overhanging leaf sees here its prototype. Internally, whether in the globe or animal body, it is a moist thick lobe, a word especially applicable to the liver and lungs and the leaves of fat. [...] The feathers and wings of birds are still drier and thinner leaves. Thus, also, you pass from the lumpish grub in the earth to the airy and fluttering butterfly. The very globe continually transcends and translates itself, and becomes winged in its orbit. Even ice begins with delicate crystal leaves, as if it had flowed into moulds which the fronds of water-plants have impressed on the watery mirror. The whole tree itself is but one leaf, and rivers are still vaster leaves whose pulp is intervening earth, and towns and cities are the ova of insects in their axils.

When the sun withdraws the sand ceases to flow, but in the morning the streams will start once more and branch and branch again into a myriad of others. You here see perchance how blood-vessels are formed. If you look closely you observe that first there pushes forward from the thawing mass a stream of softened sand with a drop-like point, like the ball of the finger, feeling its way slowly and blindly downward, until at last with more heat and moisture, as the sun gets higher, the most fluid portion, in its effort to obey the law to which the most inert also yields, separates from the latter and forms for itself a meandering channel or artery within that, in which is seen a little silvery stream glancing like lightning from one stage of pulpy leaves or branches to another, and ever and anon swallowed up in the sand. It is wonderful how rapidly yet perfectly the sand organizes itself as it flows, using the best material its mass affords to form the sharp edges of its channel. Such are the sources of rivers. In the silicious matter which the water deposits is perhaps the bony system, and in the still finer soil and organic matter the fleshy fibre or cellular tissue. What is man but a mass of thawing clay? The ball of the human finger is but a drop congealed. The fingers and toes flow to their extent from the thawing mass of the body. Who knows what the human body would expand and flow out to under a more genial heaven? Is not the hand a spreading palm leaf with its lobes and veins? The ear may be regarded, fancifully, as a lichen, umbilicaria, on the side of the head, with its lobe or drop. The lip—labium, from labor (?)—laps or lapses from the sides of the cavernous mouth. The nose is a manifest congealed drop or stalactite. The chin is a still larger drop, the confluent dripping of the face. The cheeks are a slide from the brows into the valley of the face, opposed and diffused by the cheek bones. Each rounded lobe of the vegetable leaf, too, is a thick and now loitering drop, larger or smaller; the lobes are the fingers of the leaf; and as many lobes as it has, in so many directions it tends to flow, and more heat or other genial influences would have caused it to flow yet farther.

Thus it seemed that this one hillside illustrated the principle of all the operations of Nature. The Maker of this earth but patented a leaf. What Champollion will decipher this hieroglyphic for us, that we may turn over a new leaf at last? This phenomenon is more exhilarating to me than the

> luxuriance and fertility of vineyards. True, it is somewhat excrementitious in its character, and there is no end to the heaps of liver, lights, and bowels, as if the globe were turned wrong side outward; but this suggests at least that Nature has some bowels, and there again is mother of humanity. (Thoreau 1982, 544–548; his emphases)

This region is what Mikhail Bakhtin called the grotesque bodily lower stratum and what I have called the grotesque lower earthly stratum (both discussed in the following chapter).

8.4 Wild Nature Writing

Novels, and English literature in general, for Thoreau did not have enough of what he called "wildness." Thoreau was a wild nature writer and not a tame "nature lover." Thoreau expressed in "Walking" his preference for wild thinking and literature over the tame varieties of both:

> In literature it is only the wild that attracts us. Dullness is but another name for tameness. It is the uncivilized free and wild thinking in *Hamlet* and the *Iliad*, in all the scriptures and mythologies, not learned in the schools, that delights us.[2] […] English literature, from the days of the minstrels to the Lake Poets—Chaucer and Spenser and Milton, and even Shakespeare, included—breathes no quite fresh and, in this sense, wild strain. It is an essentially tame and civilized literature, reflecting Greece and Rome. Her wilderness is a green wood, her wild man a Robin Hood. There is plenty of genial love of Nature, but not so much of Nature herself. Her chronicles inform us when her wild animals, but not when the wild man in her, became extinct.[3] (Thoreau 1982, 615–616)

English literature reproduced the ancient Greek and Roman antipathy and opposition to swamps and marshes. Ancient Greek culture thought they mixed the elements and caused disease ('malaria') (Giblett 2024, chapter 4). Ancient Roman culture thought they were not the home of agriculture, but of barbarians (Giblett 2024, chapter 5). Both thought they were the underworld populated with monsters. English literature inherited and recycled these views with a Christian twist that peopled the Fens with pagans and monsters, and justified drainage by royalty, royalists,

[2] Thoreau (1962, *II*, 97) first wrote the first three sentences in his journal.
[3] Thoreau (1962, *II*, 144) first wrote the remaining sentences in his journal.

and parliamentarians alike (Giblett 2024, chapter 6). English literature chronicles love of nature, including the life and extinction of her wild animals, but not wild men, so it separates off 'man' and nature. It does not regard 'man' as part and parcel of nature. The wild 'man' is part and parcel of nature and 'his' history and nature's should be told in nature chronicles, as Thoreau does in his own case in the first place, his place, the place of Walden.

Wild literature that chronicles the history of nature (not natural history) is expressed in the concrete language of nature, not in the abstractions of science. Thoreau (1962, *XIII*, 30) wrote in his journal, "[H]ow much better to describe your object in fresh English words rather than in those conventional Latinisms!" referring to botanical names. Thoreau asks in "Walking":

> Where is the literature which gives expression to Nature? He would be a poet who could impress the winds and streams into his service, to speak for him; who nailed words to their primitive senses, as farmers drive down stakes in the spring, which the frost has heaved; who derived his words as often as he used them—transplanted them to his page with earth adhering to their roots; whose words were so true and fresh and natural that they would appear to expand like the buds at the approach of spring, though they lay half smothered between two musty leaves in a library—aye, to bloom and bear fruit there, after their kind, annually, for the faithful reader, in sympathy with surrounding Nature. (Thoreau 1982, 616)

Thoreau preferred the primitive senses of fresh English words to the conventional elaborations of decaying Latinisms (including in English). He also wanted to keep his distance when he and his visitors "utter the big thoughts in big words" (Thoreau 1982, 391). He preferred words with traces of nature sticking to them and writing nature in wild literature, like earth adhering to the roots of a plant transplanted to the page, as they do in Thoreau's writings.

Rather than in Latinate abstractions of natural history written in prose, Thoreau sought the wild in the concrete language of poetry:

> I do not know of any poetry to quote which adequately expresses this yearning for the Wild. Approached from this side, the best poetry is tame. I do not know where to find in any literature, ancient or modern, any account which contents me of that Nature with which even I am acquainted. You

will perceive that I demand something which no Augustan nor Elizabethan age, which no *culture*, in short, can give. (Thoreau 1982, 616; his emphasis)

Thoreau demanded what wild nature can give, not tame culture.

Thoreau probably saved his most acerbic criticism for the nature lover. He wrote in his journal that "I like better the surliness with which the woodchopper speaks of his woods, handling them as indifferently as his axe, than the mealy-mouthed enthusiasm of the lover of nature" (Thoreau 1962, *I*, 237). By contrast, he found "a true lover of nature" in James Thomson, the author of *The Seasons* (Thoreau 1962, *I*, 467). Thoreau certainly loved nature and wrote nature, but he was not the affected, literary lover of nature. Thoreau was an exemplary nature writer of faction who handled nature lovingly with his quill pen in prosaic and poetic writing. He also acknowledged the grotesque lower bodily and earthly strata, especially swamps.

References

Giblett, Rod. 2023. *Middlemarsh: The Hopkins River, Kindred Wetlands and Remarkable People*. London: Transnational Press.

Giblett, Rod. 2024. *Wetland Cultures: Ancient, Traditional. Contemporary*. New York: Palgrave Macmillan.

McPhee, John. 2004. Introduction. In *A Week on the Concord and Merrimack Rivers*, ed. Henry David Thoreau, ix–xlvi. Princeton, NJ: Princeton University Press.

Solnit, Rebecca. 2014. *The Encyclopedia of Trouble and Spaciousness*. San Antonio, TX: Trinity University Press.

Thoreau, Henry David, trans. 1932. *The Transmigration of the Seven Brahmans: A Translation from the Harivansa of Langlois*. New York: William Edwin Rudge.

Thoreau, Henry David. 1958. *Consciousness in Concord: The Text of Thoreau's Hitherto 'Lost Journal' (1840–41), Together with Notes and a Commentary by Perry Miller*. Boston: Houghton Mifflin.

Thoreau, Henry David. 1962. In *The Journal 1837–1861, Volumes I–XIV*, ed. Bradford Torrey and Francis Allen. New York: Dover.

Thoreau, Henry David. 1982. In *The Portable Thoreau*, ed. Carl Bode. New York: Viking Penguin.

Thoreau, Henry David. 1995. *Walden: An Annotated Edition*, foreword and notes by Walter Harding. Boston: Houghton Mifflin.

Thoreau, Henry David. 1998. In *A Week on the Concord and Merrimack Rivers*, ed. H. Daniel Peck. New York: Penguin.

Thoreau, Henry David. 2001. *Collected Essays and Poems*. New York: Library of America.

PART II

Nature Writing and the Politics of Places and Processes

CHAPTER 9

Nature Writing and the Politics of Swamps and Marshes: Henry David Thoreau, Sidney Lanier, Aldo Leopold, and John Muir

9.1 Henry David Thoreau and Holy Dismal Swamps

Thoreau is famous as a nature writer and environmental philosopher of the wilds and the woods, an early adopter and proponent of evolutionary biology, and a pioneer of forest ecology. He is also a nature writer, environmental philosopher, and psychopolitical ecologist of bogs, marshes, and swamps. He has been canonized as the "patron saint of swamps" (Giblett 1996, ix, 229; 2020, chapter 4). All his biographers neglect or overlook this aspect of his life and work (Harding 1982; Richardson 1986; Dann 2017; Walls 2017; Buell 2024). So do both the historian of wilderness and the American mind (Nash 1982) and the philosopher of the American idea of wilderness (Oelschlaeger 1991) in their chapters on Thoreau. Wilderness for them does not include swamps, though it did for him in psychopolitical ecology.

Three Thoreauvians are the exceptions who discuss swamps in his work. Carl Bode (1982, 686) notes Thoreau's "love for swamps" and how "he enjoys being in them, enjoys writing about them." He is a paludiphile, a lover of swamps (Thoreau 2017). Branka Arsić (2016, 222–225; 385) discusses Thoreau's "philosophy of wetlands." "Swamp" for Andrew Menard (2018, 176) is "a defining figure in Thoreau's work" according to the index of *Learning from Thoreau* without elaborating much in the body of the book on what this means. The index entries refer to pages in which

R. Giblett, *The Politics of Nature and Environmental Writing*, Transdisciplinary Environmental Humanities 1,
https://doi.org/10.1007/978-3-032-04427-3_9

Menard and Thoreau mention swamps, and not to a discussion of the swamp as "a defining figure."

Annie Proulx (2022, 133–134) in her "short history" of fens, bogs, swamps, peatlands, and their destruction suggests that "Thoreau has been called the patron saint of swamps because in them he found the deepest kind of beauty and interest." This is not the reason I called him "the patron saint of swamps." It was because he upset the conventional view that privileged cities over swamps and the standard swamp-speak that denigrated swamps, and because he found sacrosanct holiness in "the darkest and, to the citizen, most dismal swamp" (Thoreau 1982, 613). He did not find "beauty" and "interest" in them, words not used by him in relation to swamps. Thoreau also found in swamps "the deepest kind" of religion of love for a place (more than the mere "fondness" for them she goes on to ascribe to him). Proulx's book does not acknowledge Thoreau's religion of love for swamps and the long western cultural history of wetlands (Giblett 1996, 2021, 2024). Her book is situated firmly on the sciences' side of the two cultures divide with the humanities.

The first part of the present chapter partly aims to address the importance of sacred swamps in Thoreau's work across the two cultures divide that did not exist for him as he was transdisciplinary before the term was invented and the divide was made. It also partly aims, not only to supply this lack or to fill this absence for its own sake but also to demonstrate the importance of Thoreau's writings for nurturing a psychopolitical wetland ecology in the country. Marshes and swamps were a persistent topic after his two-year, two-month, and two-day sojourn at Walden Pond where and when he was concerned with lakes and ponds, not wetlands. After he had left Walden Pond in 1847 and before *Walden* was published in 1854 he wrote in his journal in 1850 that "I have been surprised to discover the amount and the various kinds of life which a single shallow swamp will sustain" (Thoreau 1962, *II*, 13). He goes on to discuss a small meadow a quarter of a mile south of the pond. In later life he was not surprised to discover the biodiversity of wetlands.

Marshes and swamps were the womb and birthplace of American conservation, whereas Walden Woods and Pond were its later cradle after it had been born and where it was nurtured (Walden Woods Project 1991). When he also wrote later and lovingly about swamps, such as in "Walking," he upset and overturned mainstream thinking about swamps, including the conventions of European landscape aesthetics, philosophy, medicine, and psychology that regarded them as dismal, disease-ridden, despondent, and displeasing. In this essay and other writings, Thoreau developed a conservation counter-aesthetic of swamps, a counter-Christian religion and theology of swamps, and a psychopolitical wetland ecology of the country. He valued swamps as quaking zones where body and mind meet in the pre-modern wetland. As a proponent of evolutionary biology and as a non-believer in the institutionalized religion of Christianity, he developed in swamps his conservation counter-theology against creationism and for evolution.

Walden, his masterpiece, is largely to blame for this neglect or oversight regarding the importance of wetlands for Thoreau, especially as the first edition bore the subtitle, *or Life in the Woods.* He never wrote a book with the title and subtitle, *Concord, or Life in the Marshes and Swamps.* It is up to others to try to follow in his footsteps into the swamp and wade into marshes as he enjoined in *Walden* (Thoreau 1982, 557). Thoreau was a writer about wildness first and foremost, irrespective of whether wildness was in the dark woods, or in the dismal swamps and monstrous marshes, or on a sublime mountain, or in "the bog in our brains and our bowels" as he put it (Thoreau 1962, *IX*, 43). Wildness, as Buell (2024, 13) puts it, is "a property of mind," but it is also a property of the body and the earth. Wildness is a property of the bog in our brains, our bowels, and the earth. Thoreau is the epitome of the boggy counter-tradition to the mainstream western cultural denigration of swamps (Giblett 2021). He is the original creator of psychopolitical wetland ecology. Aldo Leopold, Sidney Lanier, and John Muir are vital contributors to psychopolitical wetland ecology as we will see later in the present chapter. More than mere nature writers, all four were psychopolitical ecologists, especially when they came to swamps and marshes.

When Thoreau lived at Walden Pond, he also lived close to Wyman Meadow, what is today described as a wetland and is signposted as such on the trail to the site of his house. Yet while he was living at Walden Pond, he did not write about much about Wyman Meadow. He certainly did not acknowledge in writing the biological richness of this wetland. Although he overlooked this wetland both literally (as he could not see it from his house) and figuratively (as he did not write about it in any detail), this proximity to a wetland may have begun to instil and nurture in him unconsciously his love for swamps that was developed and distilled in his later writings, especially in his posthumously published essay "Walking" in which the swamp is a sacred place. In this essay and other writings, Thoreau developed a counter-Judeo-Christian religion of swamps as a sacred space and a psychopolitical wetland ecology of the country.[1]

Thoreau's view of swamps shifted from being a place for bodily immersion to being a sacred site for unifying body and spirit in psychopolitical ecology. Thoreau wrote in *A Week* that "I can fancy that it would be a luxury to stand up to one's chin in some retired swamp a whole summer day, scenting the wild honeysuckle and bilberry blows, and lulled by the minstrelsy of gnats and mosquitoes!" (Thoreau 1998, 243). One could even go one step further and deeper than Thoreau did by following in the footsteps of Aldo Leopold, the patron saint of marshes, when he relates that "one day I buried myself, prone in the muck of a muskrat house. While my clothes absorbed local color, my eyes absorbed the lore of the marsh" (Leopold 1949, 160). He was eye-deep in a wetland. Leopold did not practice total immersion to the point of blocking off the sense of sight, nor Thoreau to the point of blocking off the senses of taste, smell, and hearing. Thoreau and Leopold were swamp self-baptists. Thoreau wrote in his journal that "far from being poisoned in the strong water of the swamp, it is a sort of baptism for which I had waited" (Thoreau 1962, *IX*, 376–377). Swamp baptism unifies body and spirit in psychopolitical ecology.

Thoreau not only wrote on swamps but also maintained that swamps wrote on him, or at least on his heart. He said that "when I die, you will find swamp oak written on my heart" (cited by Richardson 1993, 17). The swamp not only wrote on Thoreau; he also wrote repeatedly on it. Thoreau (1962, *II*, 13) wrote early on in his journal that "I have been surprised to discover the amount and various kinds of life which a single shallow swamp will sustain." The same could be said about Wyman Meadow. Thoreau goes on to discuss a meadow on "the south side of the pond" (and not Wyman Meadow on the north side). Two years later Thoreau was no longer surprised by the biological richness of the swamp. Thoreau (1962, *IV*, 281) wrote in his journal that the swamps are "the wildest and richest gardens that we have. Such a depth of verdure into which you sink." Thoreau was no mere walker by the wetland, but a wanderer in the wetland who was not afraid of sinking into it, provided he eventually found "a hard bottom" somewhere (as we will see later). It

[1] For further discussion of Wyman Meadow as a wetland that Thoreau overlooked, see Giblett (2025, chapter 2).

took another five years for Thoreau (1962, *X*, 196–197) to describe in his journal the plants of a swamp, as he did with Gowing's Swamp.

The richness and verdure of the swamp made it into his "real garden." For Thoreau (1993, 198; see also 2000, 32) writing in 1860–1861 "some rich withdrawn and untrodden swamp [...] is your real garden." Thoreau sought refuge and renewal in the swamp as a sacred site, indeed as the inner sanctum, the holy of holies into which, like the Jewish High Priest, he would "annually go on a pilgrimage" (Thoreau 1993, 197) and "annually we go on a pilgrimage to these sacred places" (Thoreau 2000, 31). Thoreau (1982, 612) stated in his posthumously published essay "Walking" that "I derive more of my subsistence from the swamps which surround my native town than from the cultivated gardens in the village." The swamp for Thoreau became a site that gave both physical and spiritual sustenance. Thoreau (1962, *IV*, 449) declared in his journal that "my temple is the swamp." Thoreau (1982, 613) describes in "Walking" how, "[w]hen I would recreate myself, I seek the darkest wood, the thickest and most impenetrable and to the citizen, most dismal, swamp. I enter a swamp as a sacred place, a *sanctum sanctorum* [holy of holies]." Thoreau is not merely alluding to, and countering, the Great Dismal Swamp on the border between Virginia and North Carolina, but refusing and inverting (if not subverting) the solid citizen's, and the dominant cultural paradigm's, view that all swamps are dismal by regarding them as the most sacred of places, as the holy of holies as valued by the alternative cultural paradigm and the slimy denizens of the swamp (see Giblett 2011, chapter 1 for these two cultural paradigms). Thoreau was also propounding psychopolitical ecology before the term was invented and relating it to wetlands before ecology did so too.

Thoreau found religion in the swamp, rather than at the pond, because this is where he found love. Thoreau (1958, 52; 2013, 89) wrote in a letter that "our religion is where our love is." God is not love (*contra* the New Testament Epistle of 1 John 4: 8 in which "God is love"), but religion is love, or more precisely where our love is for Thoreau. He also found religion in the swamp because this is where he found what Benjamin (1996, 69) calls "the mute magic of nature." He asked the question in his journal "what is religion?" and gave the answer "that which is never spoken" (Thoreau 1962, *XI*, 113). Religion is silent, speechless, the unspoken. God for Thoreau is not the word (*contra* the New Testament gospel of John 1: 1 in which the word is God). God for Thoreau is pre-lingual.

By contrast, God is the word and vice versa for Benjamin in his early "Language" essay (Benjamin 1996, 62–74).[2]

Rather than being merely contrarian for its own sake (as he was sometimes accused of being), religion for him meant being and living affirmatively at home in one's body, time, and place.[3] Such places for him were Walden Pond and Woods, and Concord swamps. When one finds love, one finds religion. Thoreau found love in a swamp, but also faith and hope. Thoreau (1993, vii; 2001, 442) famously had, professed, and proclaimed "great faith in a seed" and was "prepared to expect wonders" from it. He also had great faith and hope in a swamp where he was primed to experience wonders. Faith, hope, and love are the three primary fruits of the spirit according to St Paul (1 Corinthians 13:13); for Thoreau, they are the fruits of the swamp. Swamp is spirit in Thoreau's theology. As spirit is for St Paul in Christian theology, swamp is for Thoreau in psychopolitical ecology and conservation counter-theology.

Home for Thoreau is variously the body and place in the present. Hope for the future for Thoreau lay in local, uncultivated places, in the country, and not in the city. He announced in "Walking" that:

> hope and the future for me are not in lawns and cultivated fields, not in towns and cities, but in the impervious and quaking swamps. [...] I derive more of my subsistence from the swamps which surround my native town than from the cultivated gardens in the village. [...] A town is saved, not more by the righteous men in it than by the woods and swamps that surround it. (Thoreau 1982, 611–613)

Thoreau loved being immersed bodily in rivers and swamps with his senses all alive.

The opposite of hope is despair. Despair is black (as in Shakespeare's *Richard III*; see Giblett 2021, chapter 3), but black is not despair. Rather, black is hope. Hope lives in the depths of living black waters; the heights of hope spring eternal from the depths of despair in the slough of despond. The eternal now irrupts in chronological time. Thoreau deconstructs the metaphysical hierarchy of values of high over low, upper over lower, white over black, light over darkness, present over past, present over absent. He helps with the construction of the argument against the mainstream

[2] For further discussion of Benjamin's philosophy of language, see Stern (2019).

[3] For Thoreau, theology and his view of religion as being and living at home in one's body, time, and place, see Giblett (2025, chapter 10).

western denigration of wetlands and makes the case compelling for their conservation. This case is -cultural—paludicultural, agricultural, aquacultural, horticultural—in a word, ecocultural, the culture of the earthly household. The depths of hope in the slimy black waters of the swamp undermine the heights of despair in the sublime city and in the dead black waters of its slums and industrial wetwastelands. Hope springs eternal from the grotesque lower bodily and earthly strata. The grotesque lower earthly stratum is connected to and contiguous with what Mikhail Bakhtin (1984) called "the grotesque lower bodily stratum."[4] The grotesque subverts the heady heights of sublimated theory. Thoreau wrote in his journal a short and commanding prohibition: "do not despair of life" (Thoreau 1962, *X*, 228). In other words, in the affirmative, have hope in life.

For Thoreau (1982, 613), the swamp is "the strength, the marrow of Nature." Swamps nurture and nourish the marrow in the bones of the body of the earth and strengthen them. The strength of nature, for him, lies not in the hard bones of the dry land, but in the soft marrow of the marshes and swamps, what he also called the liquor of nature that feeds the earthly body, the body environmental. He wrote in his journal that "the very sight of this half-stagnant pond-hole, drying up and leaving bare mud [...] is agreeable and encouraging to behold, as if it contained the seeds of life, the liquor rather, boiled down. The foulest water will bubble purely. They speak to our blood, even these stagnant, slimy pools" (Thoreau 1962, *IV*, 102). Unlike the stagnant pools in city slums with death-dealing water that repulsed Friedrich Engels, the stagnant pools of country swamps for Thoreau speak to our blood because they contain live-giving water. Water for Thoreau (1962, *XIII*, 163) is "the most living part of nature. This is the blood of the earth." Water for Thoreau is the lifeblood of the body of the earth. Thoreau figures the earth as body, as traditional, pre-modern cultures do (discussed in a later chapter).

Thoreau's blood circulates with the blood of the earth and with the liquor and marrow of swamps in the body of the earth. Thoreau wrote in his journal and in *A Week* that "surely one may as profitably be soaked in the juices of a swamp for one day as pick his way dryshod over sand" (Thoreau 1962, *I*, 141–142; 1998, 242). Thoreau would prefer the problems of travel through the wetland, the marrow of the earth, to the ease of passage over the dry land, the bones of the earth. Yet the problems of

[4] For further discussion of the swamp as the grotesque lower earthly stratum, see Giblett (1996), and the grotesque body, see Giblett (2008a, chapter 4).

travel across the wetland are seasonal anyway in the higher latitudes as "the deep, impenetrable marsh, where the heron waded and bittern squatted [in summer], is made pervious [in winter] to our swift shoes, as if a thousand railroads had been made into it" (Thoreau 1982, 71). Thoreau sees himself as part and parcel of nature, as circulating in the body of nature, not only via the circulatory system of rivers (as in his journey on the Concord and Merrimack Rivers), but also in the semi-stagnant system of marrow through his immersion in the swamp by a pre-secular baptism.

Without the wetland, without wetlands, the world would fall apart. The wetland feeds and holds together, nurtures and coheres, the skeleton of the body of the earth. Without the wetland there would be nothing to replenish the skeletal system of the dry land, the backbones of mountain ranges, the ribs of ridges, the limbs of peninsulas and capes, and the fingers of land reaching into the sea, all of which (including the marrow of the wetlands) supply and make possible the fertile plains, prairies, and steppes on which agriculture takes place, on which the dominant cultural paradigm relies, on which industry depends, on which cities "live," or more precisely which they parasitically suck dry.

Thoreau prefers the resources of hope cultivated by the alternative cultural paradigm in the native swampy quaking zone in the country in preference to the monuments to the past enshrined by the dominant cultural paradigm in the horticultural, agricultural, architectural, industrial, and feral quaking zones in the city. Thoreau's rhetorical tactic against the stratagems of standard swamp-speak within the dominant cultural paradigm was to displace and upset the usual or normative disjunction between swamp and garden by seeing the swamp as garden, and so exploit the favourable associations of the garden as a place of light and life.[5]

Thoreau's writings on swamps touch on and counter, or subvert, many aspects of standard swamp-speak in the dominant cultural paradigm of mainstream western culture that views marshes and swamps in denigratory terms as places of darkness, disease, horror, and the monstrous—the obscenery of marshes and swamps, or "nature's shames and ills" as Charles Cotton (cited by Giblett 2011, 65) called them—that should remain hidden and not come into the open, but which do come out into the open in Thoreau's writings on swamps. Although Thoreau by no means addressed, and countered, every point of standard swamp-speak (as he seemed to be

[5] See Giblett (1996) for standard swamp-speak and Giblett (2011, chapter 1) for the dominant and alternative cultural paradigms.

more concerned to produce affirmative press for swamps and support the alternative cultural paradigm, rather than simply rebuff, or even rebut, the denigratory), he did turn the rhetoric against itself in psychopolitical wetland ecology.

Thoreau countered the prevailing miasmatic theory of disease (which maintained that one could catch diseases, such as malaria (literally "bad air"), by breathing bad air), by stating in his late essay "Huckleberries," in his journal, and in *Wild Fruits* how "miasma and infection are from within, not without" (Thoreau 1962 *V*, 394; 1980, 261; 2000, 238; 2001, 501). He upset this theory by suggesting in his early essay "A Winter Walk" how "the steam which rises from swamps and pools is as dear and domestic as that of our own kettle" (Thoreau 1982, 61). For Thoreau, swamps and stagnant pools were not the antithesis of, nor a threat to, the homely, but of comparable value. He did not valorize the wetland over the homely, but gave them equal value, unlike those of his contemporaries who denigrated and feared the wetland (and accordingly valorized the canny over the uncanny).

Rather than seeing the airs of swamps as bearers of disease, Thoreau made a crucial distinction between mist and miasma, and even saw mist as healing in a poem that acquired the title of "Mist" or "Fog." Thoreau also wrote his in journal that "the fog" is "the earth's breath made visible" (Thoreau 1962, *XI*, 404). Thoreau (1982, 237–238; 2010, 150) referred to it in a poem devoted exclusively to the subject and entitled simply "Fog" as "dull water spirit—and Protean god," as "incense of earth," "spirit of lakes and rivers," and as "night thoughts of earth." Rather than a vector for disease and a cause of death like miasma, fog is a source of new life. Rather than regarding fog and mist as vapours bearing disease and death, why not see them as the visible manifestation of the exhalations of the earth, particularly of the trees, on which we are dependent for life? After all, humans are in symbiosis with the oxygen-producing plants of the earth.

The swamp vapours were as homely for Thoreau as kettle steam because the swamp itself was better than a homely garden. Indeed, if Thoreau (1982, 612–613) had to choose between them, he would chose the swamp every time: "yes, though you may think me perverse, if it were proposed to me to dwell in the neighborhood of the most beautiful garden that ever human art contrived, or else of a Dismal Swamp, I should certainly decide for the swamp." Rather than a reason for avoiding the black swamp of despair, depression, and melancholia, Thoreau (1962, *X*, 150) suggested in his journal that "if you are afflicted with melancholy [...], go to the swamp." He prescribes and practices the walking cure for environ-mental illness and for environ-mental health. For Thoreau writing in his journal and repeating it in *Walden* "there can be no very black melancholy to him who lives in the midst of Nature and has his senses still" (Thoreau 1962, *I*, 364; 1982, 382). The place par excellence in which to live literally in the middle of Nature, even up to one's chin, is the swamp. Thoreau prescribes and practiced the swamp cure for environ-mental illness and for environ-mental health in his psychopolitical wetland ecology.

Thoreau upset the conventional view of the dominant cultural paradigm that swamps were poisonous by parodying it in his reference to the "rank and venomous luxuriance in this swamp" (Thoreau 1962, *IX*, 60). Even at the worst of times he could prescribe a swamp cure, such as when he wrote in his journal that "[w]hen life looks sandy and barren, is reduced to its lowest terms, we have no appetite, and it has no flavor, then let me visit such a swamp as this, deep and impenetrable, where the earth quakes for a rod around you at every step, with its open water where swallows skim and twitter" (Thoreau 1962, *IV*, 231). When desire is diminished and life is dissatisfying, both figured here orally, the quaking zone of the swamp has depths and softness that the shallowness of its waters belie, and that the depths of the lake cannot dream of. Thoreau values precisely those usual denigratory connotations in the dominant cultural paradigm that attach to the "depth," or horizontal extension and impenetrability of the swamp.

Thoreau also upsets the usual dissociation between swamp and city by seeing the city as swamp and so subverts the unfavourable connotations of the swamp as a place of darkness, disease, and even death. Rather than the swamp, Thoreau saw "society," "civilization," and the modern city as bearers of disease, or perhaps more precisely he saw the modern city as swamp in the conventional sense of an uncanny and unhomely place of disease and horror, and saw simultaneously the swamp as canny and homely, as postmodern dwelling in the unconventional sense of a homely, but also wild, (homely *because* wild) place. Thoreau (1962, *II*, 47) wrote in his journal that he could "see less difference between a city and some dismallest swamp then formerly. It is a swamp too dismal and dreary, however, for me." The city for Thoreau is dismal, not the swamp. He upsets and overturns the conventional association between swamp and dismalness, and the city not being dismal. Although he would prefer the swamp as swamp over the city as swamp, he nevertheless goes on to make a finer distinction: "I would prefer even a more cultivated place, free from miasma and crocodiles." And wouldn't we all? He still prefers the swamp to the city, and the native quaking zone of the swamp where crocodiles live to the feral quaking zone of the city with its swampy slums (Giblett 2009, 1–54).

Perhaps Thoreau was wise to prefer "a more cultivated place" free of crocodiles, though this did not mean that he preferred the city to the swamp, especially as the former could be as dangerous in common parlance as the latter. The city as swamp for Thoreau (1982, 350) in *Walden* had its own diseases and horrors:

> [L]et us settle ourselves, and work and wedge our feet downward through the mud and slush of opinion, and prejudice, and tradition, and delusion, and appearance, that alluvion which covers the globe, through Paris and London, through New York and Boston and Concord, through church and state, through poetry and philosophy and religion, till we come to a hard bottom and rocks in place which we call *reality*. (His emphasis)

The four cities and the town mentioned were founded in, or next to, and expanded into, marshes or swamps.[6] Thoreau is here using "the mud" as a pejorative figure for noise, or interference in the communication channel. Writing in his journal the year before *Walden* was published "the mud" is "Nature's womb" (Thoreau 1962, *V*, 499). "The mud" is an ambivalent substance, a double figure. So, let us settle ourselves, and work and wedge our feet downward through the noise of the muddy media (including social media) and through the cities built in muddy swamps, and so on, till we come to a soft bottom and mud in place which we call *reality*.[7]

9.2 Aldo Leopold, His Beloved Swamp, and the Goodness of the Undrained Marsh

Aldo Leopold (1887–1948) is regarded as "the father of wildlife conservation in America" and the author of the conservationist's "bible," *A Sand County Almanac*, or at least the conservationist's new testament to put

[6] For discussion of Paris, London, New York, and Boston, see Giblett (2016, chapters 3, 4, 9, and 10).

[7] For further discussion of Thoreau, see Giblett (2025).

beside the old testament of Thoreau's *Walden*. He has been canonized as "the patron saint of American wildlife conservation." Leopold is far better known in this role and for preaching the good news of a land ethic and conservation aesthetic in the *Almanac* than he is for writing a word for wetlands, particularly marshes, qualifying him for canonization as St Aldo of North America, the patron saint of marshes and a contributor to psychopolitical wetland ecology. Yet sprinkled throughout *A Sand County Almanac* are references to and celebrations of wetlands, even elegies for their loss, as there are elsewhere in his writings, such as in its 1947 "Foreword" not published until forty years later in which Leopold recalls autobiographically how:

> my first doubt about man [*sic*] in the role of conqueror arose when I was still in college. I came home over Christmas to find that land promoters, with the help of the Corps of Engineers, had dyked and drained my boyhood hunting grounds on the Mississippi River bottoms. The job was so complete that I could not even trace the outlines of my beloved lakes and sloughs under their new blanket of cornstalks. (Leopold 1987, 282)

Leopold bemoaned, even mourned, the loss of his slough of respond as the death of an old boyhood friend killed by civil engineering. Similarly, David Suzuki (1991, vi–vii) described how "my beloved swamp" where he used to play as a boy "is covered by an immense shopping mall and parking lot," those icons of capitalist hypermodernity.

Leopold's most powerful elegiac writing about wetlands comes in one of his "Sketches Here and There" in *A Sand County Almanac* entitled, aptly enough, "Marshland Elegy," which celebrates the life and death of Sandhill cranes. For Leopold (1949, 9) marshes may be melancholy places, not because of any despondency associated with them as such, but because "the sadness discernible in some marshes arises, perhaps, from their once having harboured cranes." Mourning for the loss of cranes is experienced by him as melancholy in his psychopolitical wetland ecology. Cranes have been killed off by drainage when "the marsh was gridironed with drainage canals" (Leopold 1949, 100). Gridironing is a very apt metaphor for the straight canals, ditches, and drains which are the devices of empire that colonize nature along with the gridplan town (see Giblett 1996, chapter 3).

Gridironing is also an apt description for the way in which the marsh became a kind of "level playing field" for the farmer and the killing fields of the crane. Leopold relates how "for them, the song of the power shovel came near being an elegy. The high priests of progress knew nothing of cranes, and cared less. What is a species more or less among engineers? What good is an undrained marsh anyhow?" (Leopold 1949, 100). This slogan could well be the motto of imperialist drainage, the "highest" drainage of capitalism to modify Lenin, scrawled on the side of the monstrous power shovel or dredger like a war machine, the kind of war fascists loved to fight, and win perhaps because it was so easy to win. The power shovel is for Leopold the marsh monster. The marsh monster is not a mythical Grendel-type figure as in the old English *Beowulf* (see Giblett 1996, 182–183; 2018, chapter 2), but the modern machines of civil engineering. The song of the power shovel drowns out the song of the earth sung by the cranes by killing their habitat and critiqued by psychopolitical wetland ecology.

Farm and marsh, it seems, cannot co-exist in modern industrial agriculture, certainly not in mutually beneficial symbiosis: "man [*sic*] cannot live by marsh alone, therefore he must needs live marshless. Progress cannot abide that farmland and marshland, wild and tame, exist in mutual toleration and harmony" (Leopold 1949, 162). 'Man' cannot live by marsh alone, but then 'he' cannot live by bread alone either, nor can 'he' live without marsh and bread; therefore, 'he' must live with farm and marsh; he cannot live marshless or farmless; therefore, 'he' must live marshful and farmful in mutual toleration, or more preferably mutual aid to use Peter Kropotkin's term from his 1914 book of this title (Kropotkin 1991). Otherwise, as Leopold (1949, 162) puts it in one of his "Sketches Here and There" in *A Sand County Almanac* devoted to the province of Manitoba (and entitled as such) in Canada, "blue lake becomes green bog, green body becomes caked mud, caked mud becomes a wheatfield," and so in turn wheatfield becomes a suburb, a shopping mall, a parking lot, a hypermodern wasteland of sterile concrete and scorching bitumen critiqued by psychopolitical ecology.

Leopold was equally scathing of what he called "alphabetical conservationists" when it comes to marshes, as he was of the drivers and backers of monstrous mechanical dredgers and shovels: "a roadless marsh is seemingly as worthless to the alphabetical conservationist as an undrained one was to the empire builders. Solitude, the one natural resource still unendowed of alphabets, is so far unrecognized as valuable only by ornithologists and cranes" (Leopold 1949, 101). An alphabetical conservationist, or "twitcher," is one who ticks off his or her sighting of a Sandhill crane or spoonbill in his or her field guide, then moves on to find the next species, oblivious to the crane's or grebe's wetland habitat, not being able to see the wood for the trees, not being able to see psychopolitical wetland ecology for birds.

Such was precisely the case when Leopold visited Clandeboye Marsh in Manitoba in 1941. Clandeboye Marsh, "northwest of Winnipeg, on the southern edge of Lake Manitoba, is part of the Delta Marsh, at 15,000 ha the largest freshwater marsh in the Canadian prairies" (Wynn 2011, xii). Manitoba indeed has, or is largely, wet prairie. Delta Marsh is what could be called a mega-marsh, vast in extent, in order to distinguish it from the much smaller and more discreet prairie pothole marshes. Delta Marsh is also a Ramsar Convention Wetland of International Importance (see Giblett 2014, 181). Leopold (1949, 158) begins his sketch of Clandeboye Marsh by bemoaning the fact that "most of us have gone blind to [...] the quality of marshes. I am reminded of this when, as a special favor, I take a visitor to Clandeboye, only to find that, to him, it is merely lonelier to look upon, and stickier to navigate, than other boggy places." Leopold's visitor takes the anthropocentric point of view of the modern navigator who wants to get through the marsh and who regards the marsh as unnavigable or as an impediment to their progress, whereas Leopold takes a bird's-eye view of point "for any pelican, duckhawk, godwit or western grebe is aware that Clandeboye is a marsh apart" because they prefer this marsh to other marshes and because they resent human intrusion as "some kind of cosmic impropriety" (Leopold 1949, 158). Leopold takes the biocentric conservationist's view that sees humans as part and parcel of the cosmos, as fellow citizens in the community of all beings, not the pinnacle of the cosmos, nor the culmination of creation, nor the conqueror of crude nature, nor the lord and sovereign ruling over the land and other beings. In this regard, Leopold was following in the footsteps of John Muir who developed a biocentric view (without naming it as such) in reflecting on his encounter with an alligator in a Florida swamp on his thousand-mile walk to the Gulf of Mexico (as we saw in a previous chapter).

The modern, anthropocentric point of view applies not only to the visitor's estimation of the marsh but also to the way one of Leopold's guests regarded the birds as he "dismissed the grebe by checking off his name in the bird list, and putting down a syllabic paraphrase of the tinkling bell: 'crick-crack,' or some such inanity" (Leopold 1949, 160). The visitor is "a twitcher" who abstracts the bird from its habitat and collects sightings in the 'cabinet of curiosities' of his bird list. He cannot see the marsh because he is too busy looking at the birds, ticking them off in his bird list, and transcribing their call. He cannot see the marsh for the birds. I encountered a few of these sort of 'birdos,' as we call them in Australia, when I visited Oak Hammock Marsh, another mega-marsh and Ramsar Convention Wetland of International Importance, north of Winnipeg, and went on a guided tour of the marshes conducted by Ducks Unlimited Canada who have their headquarters there.

Leopold not only takes the bird's point of view but also takes the muskrat's point of view when he wished he were a muskrat "eye-deep in the marsh" (Leopold 1949, 19). For Leopold, the only way to see the marsh is at its own level, not from above surveying it standing up, nor commanding it from the heights above it, nor viewing it from the air, or on the map mastering it imperially from above. Leopold even fulfilled his wish when "one day I buried myself, prone in the muck of a muskrat house. While my clothes absorbed local colour, my eyes absorbed the lore of the marsh" (Leopold 1949, 160). And presumably his nose and ears absorbed its smells and sounds too (though there are no references to either) as he was absorbed into the marsh. Leopold is a guest of the muskrat in its house. Leopold's visitor is not at home in the marsh whereas the birds and Leopold are.

This sort of bodily immersion and sensory appreciation is missing from a visit today to Oak Hammock Marsh with its boardwalks, hides, scopes, and walking tours that privilege the sense of sight in accordance with the dominant cultural paradigm, visual observation, and the marsh as bird habitat and as a place apart from modern urban living, rather than providing a much more holistic and immersive experience of the marsh itself as a place within, and vital for, modern human consciousness and life. A visit by boat into the labyrinthine waterways of Netley Marsh with a knowledgeable guide, as I was fortunate to have in Richard Gosselink, who can relate the human and natural history of the all-encompassing world and homeland it was for First Nations, is a much more Leopoldian thing to do. All people currently living in Manitoba should regard Manitoban marshes as their homeland that sustains their lives and livelihoods (see Giblett 2014, chapter 10).

The quality that Leopold valued most highly in marshes was one that required a kind of willing blindness, or at least a suspension of the sense of sight, and not of other senses, particularly of hearing that could be heightened in going to the marsh, especially if one went "too early": "to arrive too early in the marsh is an adventure in pure listening; the ear roams at will among the noises of the night, without let or hindrance from hand or eye" (Leopold 1949, 61). The sense of hearing implies immediacy with the living animal by hearing it, unlike the mastery of hand or eye over its object.

Yet the sense of smell is more immediate and intimate than the senses of sight and hearing. Leopold's does not follow in this regard in the muddy footsteps of his precursor in sensing the swamp in Thoreau, whose 14-volume *Journal* Leopold and his wife were given as a wedding present by his mother, as Thoreau certainly appreciated the aromas and fragrances of the marsh and swamp. Writing in his *Journal* Thoreau (1962, *I,* 141) could "fancy that it would be a luxury to stand up to one's chin in some retired swamp a whole summer, scenting the wild honeysuckle and bilberry blows, and lulled by the minstrelsy of gnats and mosquitoes!" Such music is the counter to the song of the power shovel: "in all swamps the hum of mosquitoes drowns this modern hum of society" (Thoreau 1962, *I*, 83). In more contemporary terms, the black noise of the creatures of the black waters of wetlands drowns out the white noise of thermodynamic and electrical industrial capitalist technology in the channels of communication (see Giblett 2008b, chapter 1).

9.3 Sidney Lanier and 'the Menstruum of Marshes'

As the patron saint of marshes, Leopold was following in the muddy footsteps of Thoreau, the patron saint of swamps, and of Sidney Lanier who, like Thoreau his near nineteenth-century contemporary, turned the rhetoric of standard swamp-speak against itself by displacing the usual pejorative associations with the swamp to the city. Lanier, like Thoreau and Leopold later, saw modern industrial capitalism and modern city life as a swamp, or at least giving rise to the mal-aria and malaise associated with swamps. Lanier (1973, 13) referred in no uncertain terms to "that universal killing ague of modern life—the fever of the unrest of trade." As an antidote to the stresses of modern life Lanier (1973, 13 and 24) prescribes life in Florida with "the smooth and glittering suavities of its lakes [...]," and its swamps, and its rivers, including the lake-like and swampy Ocklawaha River, "a river without banks."

Unlike the swamp in standard swamp-speak which emits miasma and mephitic vapours, for Lanier (1973, 21) "the glassy dream of a forest over which we sailed appeared to send up exhalations of balms and doors and stimulant pungencies." These exhalations give rise not only to a pleasant bouquet but also to a counter-rhetoric, to what Lanier (1973, 23) calls "the floating suggestions of the unutterable that come up, that sink down, that waver and sway hither and thither." From the 'depths' or shallows of the wetland the unutterable comes welling up to defy not only standard swamp-speak but language itself.

The unutterable is not the sublime about which much can and has been written, but the slime about which little has and can be written because it is slippery and cannot be grasped or defined, even as a place of death, or life, or life and death. The swamp for Lanier (1973, 29) is not a place of death and disease, but of life and death: "it is endless creation succeeded by endless oblivion." Lanier (1969, 63) addresses the marshes of Glynn in a poem of this title as "Reverend Marsh, low-crouched along the sea" as if giving birth and as "Old chemist, rapt in alchemy, /[...] The menstruum that dissolves all matter." Marshes desublimate solid earth into liquid water. The marsh, indeed the wetland in general, performs the alchemical transformation of base, or dead, or better solid matter into the liquid gold of new life (not only the black gold of swamp power, or 'gas'). Marshes desublimate solid earth into liquid water.

This menstrual process of dissolving solid matter into liquid fluid is not the sublimation of the physical into the metaphysical, nor of the sexual into the philosophical, nor of swampy into the intellectual though it can participate in the transformation of solids into gas. Nor does this process involve the application of the heat and noise of those thermodynamic technologies of modern capitalist industry, of what Lanier (1973, 14) calls "our tempting hot modern civilization."

Yet as the term "menstruum" implies, the alchemical process of the wetland is figured as the ripening of the seed within the womb fed by the placenta, so the reverend marsh is more precisely a reverend marsh *mother* rather than a reverend father. And the reverend mother gives birth by the sea crouched in a 'natural' child-birth position in that most fertile of all regions, the intertidal zone. Rather than a horrifying hell of nightmares of birth, for Lanier (1969, 64 and 47) "the marsh with flooded streams/ Glimmers, a limpid labyrinth of dreams" which is "like a lane into heaven that leads from a dream."

9.4 John Muir of the Marshes and Swamps

Leopold as the patron saint of marshes was also following in the marshy footsteps of John Muir, both of whom went into Canadian marshes. Muir, as we will see in the following chapter, is better known as the patron saint of National Parks and as 'John of the Mountains' than he is as 'John of the Marshes.' Muir wrote a word for the Canadian wetland of Holland Marsh north of Toronto now transformed into "the heart of Canada's vegetable industry." Muir's encounter with the orchid *Calypso borealis* in a Canadian marsh was, according to his standard biographer, one of "the two supreme moments of his life," the other being his meeting with Emerson, and so not his meetings with President Theodore Roosevelt, nor with the Sierra Nevadas, nor with the Californian sequoias (Wolfe 1945, 146–147; see also Fox 1981, 43). Muir's experience in a Canadian marsh in 1864 constituted a kind of conversion experience to the new religion of conservation. Later in his life, Muir (1991, 5) reflected that "*Calypso borealis* still hides in the arbor vitae swamps of Canada, and away to the southward there are a few unspoiled swamps, big ones, where miasma, snakes, alligators, like guardian angels, defend their treasures and keep them as pure as paradise." The swamps to which he refers as "away to the southward" were presumably the swamps of Florida to which, and in which, he had travelled on his thousand-mile journey to the Gulf of Mexico after his sojourn in Canada. Muir (1992) devotes one chapter of *A Thousand-Mile Walk to the Gulf* to his time in Floridian swamps and to his encounter with an alligator there, which has been seen as central for his adoption of a biocentric ethic (as we saw in a previous chapter). He devoted little space in his published work to his time in Canadian marshes, though he did in his journal and letters. Evidently in later life he was embarrassed about this period in his early life as he was dodging the draft.

In March 1864, as the American Civil War was reaching its bloody climax and drawing to its by then inevitable conclusion with the victory of the northern union, and after President Lincoln had signed an order for the conscription of 50,000 men in February, the 25-year-old Muir dodged the draft by crossing the border into Canada where he spent two years, including six months in the marshes of Ontario (Fox 1981, 42; Turner 1985, 113). Here he was, as Fox (1981, 43) puts it, "like a fugitive, like the British Army deserters he occasionally encountered in the swamps." Like them, he also sought "both refuge and nurture," as Turner (1985, 113) puts it, in the swamps in a long line of rebels and runaways holing up and finding refuge in swamps (see Giblett 1996, chapter 11). However, unlike many of these rebels and runaways, he also sought spiritual nurture and botanical specimens in the swamps, or, more precisely, he sought spiritual nurture *in* floral beauty and bounty that just happened to be in a swamp. One day in 1864 he found the rare orchid *Calypso borealis* in "a dense swamp" on "the bank of a stream"[8] where he sat down and "wept for joy" (cited by Turner 1985, 43), or "cried for joy" (Badè 1924, *I*, 121).

Muir wrote of this swamp that "land and water, life and death, beauty and deformity, seemed here to have disputed empire and all shared equally at last" (Badè 1924, *I*, 118). The swamp disputed imperial attempts to classify it as land or water, to mark it on maps, to divide it within national borders, to survey along a line of sight using optical technologies as instruments of empire through its tangled vegetation, to construct the straight lines of drain, railway, and road as vehicles of empire across and through its boggy places, to vilify it as a place of death and deformity in its smelly slime; in short, to destroy it by subjecting it to an imperial imperative which needs to own everything exclusively in accordance with the dictates of the dominant cultural paradigm and its hierarchies. The swamp resisted the dominant paradigm, disputed empire, and achieved a rapprochement between them without becoming beautiful at the time of Muir's visit, though it was beautified later as a market garden. Yet the idea of empire sharing anything equally, especially with something as useless and as ugly as a swamp, is completely inimical to it. And the idea of life and death being shared equally in the swamp's ecology is totally anathema to the death-drive of empire. Empire, not the wetland, is the orally sadistic ravening maw that acquisitively consumes everything in its path; the wetland, not the empire, is a generous and reciprocating giver of life. Empire, not the wetland, is the grim reaper, the harbinger of death. Given that the term "empire" today is often used as a code word for the US, North American wetlands dispute this empire that has all the features enumerated above. Given that these features still mark empire in its attitude and actions towards wetlands, Muir's (Leopold's and Thoreau's) writings about wetlands have obviously not been taken up much by Americans to value and protect their own wetlands. Encouraging Canadians to learn from their writings about wetlands is not a matter of cultural imperialism (yet again), but of an ongoing international and decolonizing dialogue.

[8] There is much dispute about the precise location of this marsh or swamp in Canada; see Giblett (2014, 156–157) for a discussion.

Besides refusing the logic of the imperial death-drive in the dominant cultural paradigm, Muir also refused the either/or logic of the swamp as either land or water, place of life or death, aesthetically beautiful or ugly in support of the alternative cultural paradigm and its avowal of equality. For him it was a place of both/and, a place of both land and water, life and death, beauty and ugliness. Muir bemoaned the fact that "on no subject are ideas more warped and pitiable than on death" (Badè 1924, *I*, 164; Muir 1992, 70). Against what he called "morbid death orthodoxy" (of Christianity?), he celebrated "the friendly union of life and death so apparent in Nature" (Badè 1924, *I*, 164–165 and Muir 1992, 70), not least, and perhaps most powerfully, to be found in the wetland. In nature for Muir there was to be found "the beautiful blendings and communions of death and life, their joyous inseparable unity," as he found in the swamps of Canada (Badè 1924, *I*, 165; Muir 1992, 70).

If one feature could be said to characterize the difference between standard swamp-speak in the dominant cultural paradigm that denigrates wetlands and the counter-tradition in which Muir, Leopold, Thoreau, and others participate within the alternative cultural paradigm, it would be that the latter emphasize that the wetland is a place of both life and death, whereas in standard swamp-speak it is only a place of death (for the two paradigms, see Giblett 2011, chapter 1). As Muir's first biographer says, "[I]n the eyes of other people the spirit of death seemed to brood over these swamps, but to him they were reservoirs of life" (Wolfe 1945, 93). Indeed, for Muir (1987, 229, 230) "everything in Nature called destruction must be creation [...] everything is flowing" and "there are no harsh dividing lines in nature," not between life and death, nor between land and water. Like Thoreau, Muir believed that nature loves gradation, not just in colour, nor only between land and water, but even between life and death.

In the eyes of pioneering colonists, Holland Marsh may have been "a vast and useless swamp," but to the young Muir it contained the exquisite and enchanting orchid *Calypso borealis*. Also in the eyes of the young Muir, land and water "have disputed empire" in the Holland Marsh, and they did at the time as they were, as Worster (2008, 94) puts it, "a vast—and in the eyes of pioneers, a useless—swamp," but are "now drained and intensively farmed." It took little less than 80 years after Muir set foot in the Holland Marsh for it to go from refuge and sanctuary for uncommodified spiritual nurture to market gardens for commodified physical nurture and a source of profit for market gardeners in the service of the dominant cultural paradigm. Psychopolitical wetland ecology critiques, refuses, and resists the dominant cultural paradigm, advances the alternative cultural paradigm, and celebrates the life of wetlands. The nature writing of Henry David Thoreau, Aldo Leopold, Sidney Lanier, and John Muir is at the forefront of psychopolitical wetland ecology.

References

Arsić, Branka. 2016. *Bird Relics: Grief and Vitalism in Thoreau*. Cambridge, MA: Harvard University Press.

Badè, William. 1924. *The Life and Letters of John Muir, Volumes I and II*. Boston: Houghton Mifflin.

Bakhtin, M.M. 1984. *Rabelais and His World*. Translated by H. Iswolsky. Bloomington, IN: Indiana University Press. First published in Russian in 1965.

Benjamin, Walter. 1996. In *Selected Writings: Volume 1, 1913–1926*, ed. Marcus Bullock and Michael Jennings. Cambridge, MA: Belknap Press of Harvard University Press.

Bode, Carl. 1982. Epilogue by the Editor. In *The Portable Thoreau*, ed. Carl Bode, 683–696. New York: Viking Penguin.

Buell, Lawrence. 2024. *Henry David Thoreau: Thinking Disobediently*. New York: Oxford University Press.

Dann, Kevin. 2017. *Expect Great Things: The Life and Search of Henry David Thoreau*. New York: TarcherPerigree.

Fox, Stephen. 1981. *John Muir and His Legacy: The American Conservation Movement*. Boston: Little, Brown.

Giblett, Rod. 1996. *Postmodern Wetlands: Culture History Ecology*. Edinburgh: Edinburgh University Press.

Giblett, Rod. 2008a. *The Body of Nature and Culture*. Basingstoke: Palgrave Macmillan.

Giblett, Rod. 2008b. *Sublime Communication Technologies*. Basingstoke: Palgrave Macmillan.

Giblett, Rod. 2009. *Landscapes of Culture and Nature*. Basingstoke: Palgrave Macmillan.

Giblett, Rod. 2011. *People and Places of Nature and Culture*. Bristol: Intellect Books.

Giblett, Rod. 2014. *Canadian Wetlands: Places and People*. Bristol: Intellect Books.

Giblett, Rod. 2016. *Cities and Wetlands: The Return of the Repressed in Nature and Culture*. London: Bloomsbury.

Giblett, Rod. 2018. *Environmental Humanities and Theologies: Ecoculture, Literature and the Bible*. London: Routledge.

Giblett, Rod. 2020. *New Lives of the Saints: Twelve Environmental Apostles.* Lanham, MD: Hamilton Books.

Giblett, Rod. 2021. *Wetlands and Western Cultures: Denigration to Conservation.* Lanham, MD: Lexington Books.

Giblett, Rod. 2024. *Wetland Cultures: Ancient, Traditional, Contemporary.* New York: Palgrave Macmillan.

Giblett, Rod. 2025. *Waking to Nature with Thoreau and Benjamin: Or, Psychopolitical Ecology.* New York: Bloomsbury.

Harding, Walter. 1982. *The Days of Henry Thoreau: A Biography.* New York: Dover.

Kropotkin, Peter. 1991. *Mutual Aid: A Factor in Evolution.* Montreal: Black Rose. First published 1914.

Lanier, Sidney. 1969. *Poems and Letters.* Baltimore, MD: The Johns Hopkins University Press.

Lanier, Sidney. 1973. *Florida: Its Scenery, Climate, and History.* Gainesville, FL: University of Florida Press. First published 1875.

Leopold, Aldo. 1949. *A Sand County Almanac and Sketches Here and There.* Oxford: Oxford University Press.

Leopold, Aldo. 1987. 1947. Foreword. In *Companion to a Sand County Almanac: Interpretive and Critical Essays*, ed. J. Baird Callicott, 281–288. Madison: The University of Wisconsin Press.

Menard, Andrew. 2018. *Learning from Thoreau.* Athens, GA: University of Georgia Press.

Muir, John. 1987. *My First Summer in the Sierras.* New York: Penguin. First published 1911.

Muir, John. 1991. *Our National Parks.* San Francisco: Sierra Club Books. First published 1901.

Muir, John. 1992. *A Thousand-Mile Walk to the Gulf.* New York: Penguin. First published 1916.

Nash, Roderick. 1982. *Wilderness and the American Mind.* 3rd ed. New Haven: Yale University Press.

Oelschlaeger, Max. 1991. *The Idea of Wilderness: From Prehistory to the Age of Ecology.* New Haven: Yale University Press.

Proulx, Annie. 2022. *Fen, Bog and Swamp: A Short History of Peatland Destruction and its Role in the Climate Crisis.* New York: Scribner.

Richardson, Jr, and D. Robert. 1986. *Henry Thoreau: A Life of the Mind.* Berkeley, CA: University of California Press.

Richardson, Robert D., Jr. 1993. Introduction: Thoreau's Broken Task. In *Faith in a Seed*, ed. Bradley Dean, 3–17. Washington, D.C.: Island Press/Covelo, California: Shearwater Books.

Stern, Alexander. 2019. *The Fall of Language: Benjamin and Wittgenstein on Meaning.* Cambridge, MA: Harvard University Press.

Suzuki, David. 1991. Foreword. In *Taming the Great South Land: A History of the Conquest of Nature in Australia*, ed. William Lines, vi–vii. North Sydney: Allen and Unwin.

The Walden Woods Project. 1991. *Walden Woods: Cradle of American Conservation.* Boston: The Walden Woods Project.

Thoreau, Henry David. 1958. In *The Correspondence*, ed. Walter Harding and Carl Bode. New York: New York University Press.

Thoreau, Henry David. 1962. In *The Journal 1837–1861, Volumes I–XIV*, ed. Bradford Torrey and Francis Allen. New York: Dover.

Thoreau, Henry David. 1980. *Natural History Essays.* Salt Lake City, UT: Peregrine Smith.

Thoreau, Henry David. 1982. In *The Portable Thoreau*, ed. Carl Bode. New York: Viking Penguin.

Thoreau, Henry David. 1993. In *Faith in a Seed*, ed. Bradley Dean. Washington, D.C.: Island Press/Covelo, California: Shearwater Books.

Thoreau, Henry David. 1998. In *A Week on the Concord and Merrimack Rivers*, ed. H. Daniel Peck. New York: Penguin. First published in 1849.

Thoreau, Henry David. 2000. In *Wild Fruits: Thoreau's Rediscovered Last Manuscript*, ed. Bradley Dean. New York: Norton.

Thoreau, Henry David. 2001. *Collected Essays and Poems.* New York: Library of America.

Thoreau, Henry David. 2010. *Collected Poems*, enlarged edition, ed. Carl Bode. Baltimore, MD: Johns Hopkins University Press.

Thoreau, Henry David. 2013. In *The Correspondence: Volume 1: 1834–1848*, ed. Robert Hudspeth. Princeton, NJ: Princeton University Press.

Thoreau, Henry David. 2017. "The Swamp Lover: An Excerpt from his May 1862 Article 'Walking'." The Atlantic Monthly. Accessed online October 17, 2023. https://www.theatlantic.com/magazine/archive/2017/11/the-swamplover/541356/.

Turner, Frederick. 1985. *Rediscovering America: John Muir in His Time and Ours.* New York: Viking.

Walls, Laura Dassow. 2017. *Henry David Thoreau: A Life.* Chicago, IL: University of Chicago Press.

Wolfe, Linnie. 1945. *Son of the Wilderness.* New York: Alfred A. Knopf.

Worster, Donald. 2008. *A Passion for Nature: The Life of John Muir.* Oxford: Oxford University Press.

Wynn, Graeme. 2011. Foreword. In *Wet Prairie: People, Land, and Water in Agricultural Manitoba*, ed. Shannon Stunden Bower, xi–xx. Vancouver: University of British Columbia Press.

CHAPTER 10

Nature Writing and the Politics of National Parks and the Sublime: John Muir

John Muir was born in Scotland in 1838 and died in California in 1914. He spent his childhood in Wisconsin, some of his early adulthood in Canada dodging the draft for the Civil War and converting to conservation in a marsh and in Florida in a swamp encountering an alligator and converting further to a biocentric ethic, and most of his adulthood in California climbing trees and mountains, writing about them, and calling for their conservation. He was, and still is, a monumental figure in the American conservation movement, if not its patriarch. He has been described as "the spiritual father of American environmentalism" (Worster 1993, 189), as "the father of our [United States'] national parks and forest reservations" (Tilden 1970, 537), and as "a patriarch of the National Park system" (Mitchell 1994, 52). The father of the idea for national parks is generally recognized as George Catlin who is credited with proposing it in 1832 (see Giblett 2011, chapter 8). Muir has been canonized as St John the Baptist of the mountains, the patron saint of national parks. Muir was following in the footsteps of the old testament prophet of conservation of Thoreau in *Walden*. Muir was also a voice crying in the wilderness preparing the way for Aldo Leopold's saving gospel of a land ethic and conservation aesthetic in the new testament of *A Sand County Almanac*.

Muir's ideas about national parks and the gender politics of his views of nature are paradoxical. For Muir nature in general is sacralized, yet his views on where and how nature is sacralized depend on the aspect or site

R. Giblett, *The Politics of Nature and Environmental Writing*, Transdisciplinary Environmental Humanities 1,
https://doi.org/10.1007/978-3-032-04427-3_10

under discussion. For Muir nature in mountains and forests was aestheticized and sublimated, especially if such areas were 'set aside' and preserved in the sanctuaries or 'modern cathedrals' of national parks, whereas nature for him could also be inhabited, invigorated, and enlivened by all living things, including rock and stone, and even human beings (but not indigenous peoples).

Under Muir's leadership the American conservation movement has largely been concerned with conserving sanctuaries, those special places for plants and animals and those aesthetic landscapes for taking refuge from the depredations of modern urban industrial capitalism. American conservationism has often been more precisely 'sanctuarism.' It has been concerned with giving sanctuary and preserving sanctuaries, rather than with caring for the earthly household as a whole. Certainly, those special places need to be conserved—humans need sanctuaries and sanctuaries are needed for plants and animals. In the famous John Muir-Gifford Pinchot struggle over the damming of Hetch Hetchy valley in Yosemite National Park, and its many descendants, I am on Muir's and his children's side every time.

The American conservation movement arose in the aftermath of what Stephanie Coontz (cited by Flichy 1995, 67) calls "the apogee of the private sphere" (1870–1890) in which "refuge" and "sanctuary" were repeatedly and routinely applied to the home. By taking these terms from the private sphere and applying them to the biosphere, the latter was privatized. Even public lands were construed in private terms. The commons, owned by none and shared by all, were colonized in public lands, such as National Parks, and enclosed in private terms, such as "refuge" and "sanctuary."

This care for special places needs to be extended to the earth household as a whole, as a special space with special places. Deconstructing and decolonizing Muir's work with its romantic aesthetic of the sublime as a naturalistic theology and valuing his work on nature as living being is one step to doing so. It is also part of psychopolitical ecology that promotes and exercises earthly mutuality—and not a reactionary sanctuarism constantly trying to save endangered species or special places threatened with destruction. Rather than always fighting rear-guard actions, psychopolitical ecologists are the avant-garde. Political ecocriticism of Muir's nature writing advances psychopolitical ecology.

Muir sublimated mountains and forests into a patriarchal God not so much in what Murray Bookchin (1991, il) calls his "inverted Calvinism" (as the relationship between God and nature is not merely turned upside down) as his naturalistic Calvinism, or perhaps more precisely naturalistic Campbellism.[1] Instead of "the natural world as a path to a knowledge of God" (Grove 1995, 15) as in Calvinism, the natural world for Muir was a manifestation of God. Although he is often regarded as a pantheist, Michael Zimmerman (1994, 385, n.48) argues that "Muir may have been a *panentheist*, one who regards nature as a manifestation of but not exhaustive of the divine." Muir's God was incarnated in His son of the elected, and even predestined,[2] elements and features of mountains and forests such as geysers and glaciers especially, but also in rock and stone, wood and water, plants and animals, and even human beings (though again not indigenous peoples).

Yet when Muir celebrates living nature not just in forests and mountains, but also in glaciers and swamps, at work destroying and creating life, nature is thereby enlivened or invigorated in "panentheism" (see Sheldrake 1991, 198). Muir here sacralizes nature as "Mother Earth" or "Mother Nature."[3] Muir vacillates between nature aestheticized (especially sublimated and incarnated) for nature lovers, nature preserved (especially in the sanctuaries of the modern cathedrals of national parks) for city dwellers and tourists, and nature inhabited (especially enlivened or invigorated by all things including human visitors to national parks, but not for Muir by indigenous human inhabitants or native title holders of national parks).

10.1 Cathedrals of the Modern World

In a series of articles published in *The Atlantic Monthly* during the 1890s and later collected in his book, *Our National Parks*, first published in 1901 Muir expressed his ideas and views about national parks in particular and nature in general. Muir also wrote specifically and lovingly of the first

[1] For Muir's Calvinist upbringing, see Fox (1981, 28–30). See Worster (1993, 190–196) for a useful discussion of Muir's more specifically Campbellite upbringing. Worster bemoans the fact that Fox (1981) and Cohen (1984) do not even mention Campbellism and that Turner (1985, 192–193) only treats it in passing. See also Worster (2008, 37–39).

[2] According to Worster (1993, 240, n.20), the Calvinist doctrine of predestination would have been elitist for the Campbellite Daniel Muir, John's father, though John saw glaciers as predestined (as we will see below).

[3] 'Mother Earth' for Muir is not strictly speaking the Neolithic earth mother of dryland agriculture and patriarchy, nor is it explicitly the Palaeolithic wetland Great Goddess of 'matriarchy.'

two national parks in the US (and the world): Yellowstone National Park and Yosemite National Park. For Muir national parks, or more precisely forested and/or mountainous national parks, were seen as fulfilling an important inspirational role as modern cathedrals. Yet forested and/or mountainous national parks are more precisely America's outdoor modern cathedrals distinct, as we will see below, from the indoor modern cathedrals of the cinemas of industrial capitalism and the museums of the nation-state.[4]

Muir (1980, 127), however, goes one step further than simply regarding Yosemite National Park as an outdoor cathedral as he found "the mighty wilderness of mountains" of the High Sierra not only like "some Gothic cathedral" but also "more abundantly spired than Milan's." The inspiring High Sierran cathedral 'outspired' the smaller, less abundantly spired Christian cathedral. Muir was not alone amongst his conservationist predecessors and his contemporaries in this view of forested and mountainous regions. In 1858 Henry David Thoreau (1962, *XI*, 353) described in his *Journal* "the quiet and somewhat sombre aisles of a forest cathedral." Yosemite was regarded widely as "a natural cathedral" in the nineteenth century (see Sears 1989, 122 and 138). Indeed, one of the mountains of Yosemite is named Cathedral Peak which Muir (1987, 198) described as "a majestic temple [...] adorned with spires and pinnacles in regular cathedral style."

Yet Yosemite was not a Christian cathedral. For Muir (1987, 146, 198) it was "one of Nature's cathedrals, hewn from the living rock" and Cathedral Peak was "hewn from the living rock." It was left in situ unlike the cathedrals of Europe built of dead stone, the extracted and hewn resource, the transported and erected object. Nature for Muir was living agent, not dead matter. No Sierra landscape that Muir (1987, 157) saw "holds anything dead or dull, or any trace of what in manufactories is called rubbish or waste." Mountainous landscape may have been wasteland in the eyes of his contemporaries, but for Muir (1987, 60) it was where "God himself seems to be always doing his best here, working like a man in a glow of enthusiasm." God's Promethean, thermodynamic and industrial workshop, and home, for Muir is up in the mountains, not down on the lowlands in the cities, the home of citizens, and certainly not even further down in the nether lands of the swamps, the home of the

[4]Following in the footsteps of Paul Virilio, I discuss cinemas as cathedrals of industrial capitalism in Giblett (2008, chapter 5).

matrifocal Great Goddess (see Giblett 1996, especially chapter 6). As Muir (1987, 205) refers later to "Nature's glacial workshops," God is Nature and Nature God for Muir.

Despite, or perhaps because of being God's best work, mountainous landscape did not fall into the conventional aesthetic category and experience of the picturesque. For William Gilpin (1794, 43) in the late eighteenth century, "the spiry pinnacles of the mountain, and the castle-like arrangement of the rock, give no peculiar pleasure to the picturesque eye." Similarly, a hundred years later Muir (1980, 104) claimed that "to artists, few portions of the High Sierra are, strictly speaking, picturesque. The whole massive uplift of the range is one great picture, not clearly divisible into smaller ones." In other words, the High Sierra with its phallic "massive uplift" invokes the experience of the sublime as do "the ineffably chaste and spiritual heights of the Fairweather Range [in Alaska and Canada ...,] the whole making a picture of icy wildness unspeakably pure and sublime" (Muir 1980, 14).

Muir enacts what Belden Lane (1988, 22) calls "the sublime as the monumental," or perhaps more precisely, the sublime as the *experience* (the sublime is a *state* of aesthetic sensibility) of the *mountainous* (as distinct from the urban or built) monumental.[5] Again Muir was not alone in the nineteenth century in valorizing the mountainous as evocative of the sublime. Indeed, it was a commonplace of nineteenth-century landscape aesthetics (e.g., Wordsworth; see Giblett 2011, chapter 3). For Thoreau (1980, 44) too, "sublimity and grandeur [...] belong to mountain scenery" and for John Ruskin (cited by Schama 1995, 513) also "we take our idea of fearfulness and sublimity from the mountains and the sea," though he prefers the former to the latter.

Yet Yosemite was "that sublime Sierra temple" not only because it was mountainous but also because it was forested. Muir (1991, 60; see also p. 226) describes how "the sunbeams streaming through their [the pines'] feathery arches brighten the ground, and you walk beneath the radiant ceiling in devout subdued mood, as if you were in a grand cathedral with mellow light sifting through colored windows, while the flowery pillared aisles open enchanting vistas in every direction."[6] For Muir, "reservations

[5] For mountains as monuments for Muir, see Muir (1980, 35) and for monumentalism as the impetus behind the national park idea, see Runte (1979, 28, 49, and 60).

[6] For Yosemite as sublime, see Muir (1991, 67 and 70; 1987, especially chapter 5, 'The Yosemite,' where he also repeatedly refers to the Yosemite temple; and 1980, 97). Other

of scenery," as for his compatriot and near contemporary Charles Eliot (cited by Thomas 1984, 269) writing in 1896, were "the cathedrals of the modern world." Or perhaps more precisely, *mountainous forest* reservations (the forerunners of national parks) of "scenery," the habitual visual aesthetic of the sightseer and tourist, were modern cathedrals.

Yet the cathedral is very much an alien and imported construct imposed on an indigenous inhabitation of the land, the Christian (or in Muir's case his naturalistic Campbellism [but what of his panentheism here?]) proselytizing the 'pagan' and sublimating the chthonic. Couple the cathedral with the sublime in the monumental mountains and then set it aside in the sanctuary of the national park and the result is a neoromantic sacred sight (not site). In Muir's (1991, 69) naturalistic Campbellism a stern and craggy, patriarchal and mountainous triune God "brood[s like the Holy Spirit] outspread over the predestined landscape" of the glacial mountains and valleys, instead of a stern and craggy, patriarchal and monumental God brooding spirit-like over predestined saved souls of Calvinism or over the primordial wetlandscape of the littoral, lacustrine, and lagoonine as in the Biblical creation story (for a rereading of *Genesis* 1:2 in these terms, see Giblett 2018, chapter one).

By referring to "the wombs of the ancient glaciers" Muir (1980, 222) displaces the beginnings of the world from the littoral to the glacial, from the world below to the world above. His patriarchal appropriation of maternal functions by birthing metaphors and his displacement of the primordial landscape from the littoral to the glacial which does not acknowledge the primordial womb of the wetland are both crucial for his sublimatory nature aesthetic and religion. In Muir's naturalistic Campbellism, the patriarchal God-function is sublimated into the monumental mountains where God the Father Law gives birth to new life with His own labour alone in the simulated womb of His glacial workshop. The matrifocal Great Goddess of the swamps is displaced upwards and patriarchally simulated into what Muir (1980, 102 and 46) calls "glacier wombs" and "the broad white bosoms of the glaciers." The glacial white bosom is a kind of *tabula rasa*, or in Sigmund Freud's terms a mystic writing pad, on which Muir could inscribe his writings and in Freud's terms a screen on

forests could also be sublime cathedrals as in Alaska where Muir (1980, 48–49) was 'charmed with the majestic beauty and grandeur of the trees, as well as with the solemn stillness and the beauty of the elastic carpet of golden mosses flecked and barred with the sun-beams that sift through the leafy ceiling.'

which he could project his phantasies, his daydreams. The whiteness of the bosom serves to smother any trace of the living black waters of the Great Goddess of the swamps, and to sanitize any messiness in parturition.

The forested and/or mountainous national park as modern cathedral is the outdoor extra-urban counterpart to the indoor urban cathedrals of the modern world. These are what David Day (1996, 182) calls "the new stone-built cathedrals of knowledge, the public libraries, art galleries and museums" and what Paul Virilio (1989, 38) calls "the cinema-cathedrals of the modern state." Or perhaps more precisely, the cinema-cathedrals of the modern imperialist nation-state in hegemonic alliance with industrial capitalism. As with film, which was for Walter Benjamin (1986, 55) "one of the most advanced machines for the imperialist domination of the masses," the national park is a machine for the imperialist domination of the earth. "The empire of nature," as John MacKenzie (1988) calls it, is applicable not just to the nineteenth century in the colonies, but to the capitalist commodification of nature today in national parks, zoos, and other nature theme parks, as well as in nature documentaries, so-called natural products (an oxymoron), and tourism.

Yet the national park and the cinema as modern, imperialist cathedrals performed different quasi-religious functions. The early cinemas, Virilio (1989, 31) argues, were regarded as "deconsecrated sanctuaries in which, as Paul Morand put it, the public sensed the end of the world in an ambience of profanation and black masses." For Muir, by contrast, Yosemite and Yellowstone National Parks were reconsecrated sanctuaries in which 'the public' could sense the evolutionary, albeit glacial, beginning, and even ongoing recreation, of the world in an ambience of reverence and sublime 'white' masses (unlike the mephitic and miasmatic atmosphere of the slimy, living black morasses of wetlands (see Giblett 1996)). Yosemite National Park was a paradise for Muir (1991, 76) "that makes even the loss of Eden seem insignificant." The Park contained parks which were as "fair as Eden" (Muir 1991, 129), though perhaps Yosemite National Park was more what Roderick Neumann (1995, 155) has called in relation to national parks in Africa an "ersatz Eden."

Yet like the cinema with its modern technological simulation of the cathedral's stained-glass windows in which, as Virilio (1989, 38) puts it, "everything visible appears to us in the light," "the standard perception of national parks" was, as Alfred Runte (1979, 64) describes it, "as a unique visual experience" centred on the aesthetically legitimated sense of sight. Like Wordsworth, Muir (1991, 2) bemoans the fact that national parks were subjected to what he calls "the scenery habit" common amongst his contemporaries (see Giblett 2011, chapters 3 and 4). Like Wordsworth too, he was prone to what could be called the sublimatory habit, especially in relation to *mountainous* national parks. The sublimatory habit transcended the common scenery habit and placed the poet and preservationist above his contemporaries as an arbiter of high cultural taste in landscapes.

Yet by functioning as reconsecrated modern cathedrals *forested* national parks could reverse history, or at least hark back to an earlier era, by returning the cathedral to the sacred, pre–Christian, pre-patriarchal grove from which it sprang. The cathedral's "soaring columns and vaults," as Rupert Sheldrake (1991, 44–45 and 176) points out, "recall sacred groves, and vegetation bursts out everywhere. [... M]any churches and cathedrals in Europe were built on pre-Christian sacred sites." The forested national park as cathedral not only re-appropriates the appropriated sacred grove but also could return the sacred place to its pre-architectural, pre-Christian wild(er)ness manifestation, to what Devereux Butcher (1969, 1) calls "temples not built with hands" (though national parks could contain areas made and worked by indigenous hands), unlike "the great cinema temples," as Virilio (1989, 31) calls them, built with hands.

Yet like the cinema temples, Muir (cited by Cohen 1984, 192) found that "the money-changers were in the temple" of Yosemite. The forested and mountainous national park as temple needed its modern-day Jesus who would drive out the moneychangers, the cash-spinners, and usurers. One such Jesus was Muir (1991, 209; see also 225) who did not need to be reminded that "great trees and groves used to be venerated as sacred monuments and halls of council and worship" and "the hills and groves were God's first temples" (Muir 1987, 146). Muir regarded Sequoia forests in particular as "majestic living temples, the grandest of Gothic cathedrals" (cited by Cohen 1984, 195).

The forested national park as cathedral and temple has meant that, as Sheldrake (1991, 177) goes on to suggest, "a religious experience of the wilderness has endowed many of the American national parks with a transcendental quality. For many who visit them, they are more than recreational areas; they are natural temples or sanctuaries." Muir has been ordained as prophet and evangelist of this new religion of naturalistic Campbellism, or simply sanctuarism: a new St John, writer of a new gospel; a new good news about the saving graces of wild(er)ness sanctuary; a new John the Baptist, crying out in and for the wild(er)ness preparing the way for Aldo Leopold and his saving gospel of a land ethic and a conservation aesthetic; a new St John the Divine proclaiming the apocalypse of a blighted and benighted urban 'civilization' bereft of wild(er)ness; and a new John Calvin converting God/Nature dualism into the craggy patriarchal sublime and predestined glaciers of what Muir (1991, 173) called "the stern wilderness" of the mountains, and of the mountains alone. Muir even saw himself as John the Baptist: "Heaven knows that John the Baptist was not more eager to get all his fellow sinners into the Jordan than I to baptize mine in the beauty of God's mountains" (cited by Cohen 1984, 259; see also 254). Worster (1993, 194–195) also argues that Muir "became a kind of frontier evangelist" and "invented a new kind of frontier religion."

There is some support for these views of Muir in his own writings on national parks. For Muir (1991, 31), the cathedral of the national park with forests or geysers can produce a contemplative mood in "the calmest, stillest scenery" of its forests, or can render the onlooker "awe-stricken and silent, in devout, worshipping wonder" in the face of the "awful uproar" and "tremendous outburst" of its terrifying, sublime, phallicized geysers (Muir 1991, 40–41). For Muir (1979, 350), the forested national park is a public place of community worship, not a private cloister for individual meditation, or more precisely it is a place for "both solitude and society." Writing of Yosemite National Park, Muir (1991, 59) described how "nearly all the park is a profound solitude. Yet it is full of charming company, full of God's thoughts, a place of peace and safety amid the most exalted grandeur and eager enthusiastic action, a new song, a place of beginnings abounding in first lessons on life." The monumental and the mountainous again appropriate the beginnings of life.

Forested national parks are also sacred places for Muir because they are a refuge or sanctuary, like the medieval monastery, for the fugitive. For Muir (1979, 350), Yosemite National Park is "a place of rest, a refuge from the roar and dust and weary, nervous, wasting work of the lowlands," unlike the uplifting, sublimating highlands (Muir was, of course, of Scottish descent and was born in Scotland). In an increasingly urbanized world where it seems the only possible other refuge from "rust and disease" (Muir 1991, 1) is the golf course, but whose membership fees are so prohibitively expensive for most (the golf course is the rich man's private refuge), the forested national park is for Muir (1979, 352) "the poor man's [public] refuge," but not for him the indigenes' working land. Muir was very much a proponent of the sanctuary idea for national parks that would provide a refuge for native flora and fauna, and for the poor lowlanders, or city dwellers, but not a site of inhabitation and enjoyment of ownership and rights by indigenous peoples.

Perhaps the poorest 'man' culturally and spiritually of all in the modern and hypermodern worlds is the tourist, or at least more precisely the modern mass-packaged touristic consumer. For Muir (1991, 9), "the wildest health and pleasure ground accessible and available to tourists seeking escape from care and dust and early death are the parks and reservations of the West." National parks can provide the tourist with an escape route from the prison of modern city life provided once s/he gets there s/he slows down. Muir (1991, 42) rails that "nothing can be done well at a speed of forty miles *a day* [my emphasis]. The multitude of mixed, novel impressions rapidly piled on one another make only a dreamy, bewildering, swirling blur, most of which is unrememberable." If the tourist treats the national park like the city with its flood of impressions, the result will be the same, undistinguished, unremarkable blur of experience.

The tourist also needs to experience the national park alone. For Muir (1991, 210), "little [...] is to be learned in confused, hurried tourist trips, spending only a poor noisy hour in a branded grove with a guide. You should go looking and listening alone on long walks through the wild forests and groves in all the seasons of the year." The sublime is a solitary experience. A national park is a cathedral because it takes time to appreciate its vastness, its fastnesses, its nooks and crannies. In Yosemite National Park, for Muir (1991, 58), "nowhere will you see the majestic operations of nature more clearly revealed beside the frailest, most gentle and peaceful things." For Muir (1991, 30) Yosemite National Park is not only a means to escape from an early death, but also "the scenery is wild enough to awaken the dead." The scenery can even shake the apathetic out of their lethargy. The geysers and hot springs of Yellowstone National Park display "an exuberance of color and strange motion and energy calculated to surprise and frighten, charm and shake up the least sensitive out of apathy into newness of life" (Muir 1991, 31). Geysers for Muir are terrifying, sublime agents.

Geysers are also masculinized by Muir. The sublime in the masculinist modern western tradition has been associated inextricably with the masculine virtue of what Terry Eagleton (1990, 54) calls "virile strenuousness." The sublime is, as Eagleton goes on to argue, "on the side of enterprise, rivalry and individuation" and is "the lawless masculine force which violates yet perpetually renews the feminine enclosure of beauty." For Muir

(1991, 214; see also 228), for example, the "Big Tree" of the Sequoia shows "Nature's immortal virility." Similarly, geysers for Muir (1991, 31) are "like inverted waterfalls," but also like the upright mountain in that they are phallic as Muir (1991, 31–32) shortly leaves his reader in no doubt as he describes the geysers of Yellowstone National Park as "standing rigid and erect, hissing, throbbing, booming."

Muir (1991, 33) generally prefers geysers to waterfalls, both to swamps, and mountains over all three. For Muir (1991, 4) geysers are "the laboratories and kitchens, in which, amid a thousand retorts and pots, we may see Nature at work as chemist or cook, cunningly compounding an infinite variety of mineral messes." Nature is here figured ambiguously, even androgynously, as chemist or cook, even alchemist or witch, working in the laboratory and kitchen with thermodynamic technology in "the hot underworld" as God works in His glacial workshop, the industrial inferno to produce meals (me(t)als?) for modernity. Nature as figured in these terms is not all that far away from the culture of what Muir (1991, 2) calls "these hot, dim, strenuous times," these modern times in which base matter is sublimated into ethereal commodities using heat (though they are also "these cold, doubting, questioning, scientific times" (Muir 1991, 41) produced by the penetrating light of reason).

10.2 God's National Parks

Besides a view of nature sublimated in national parks, Muir also propounded a view of God-the-Father-nature incarnated in mountains, or mountainous scenery, with its dawn lights and alpen-glows as a kind of God-the-Son figure. For Muir (1991, 56) "next to the light of the dawn on the high mountain tops, the alpenglow is the most impressive of all the manifestations of God." The alpenglow, sublimated above the mountains themselves, could manifest God, and mountains could even be God. When Muir (1980, 18–19) encounters the Fairweather Mountains in Alaska he regards them as God Himself: "here the mountains themselves were made divine." For Muir, the mountains *were* God.

If the mountains could be God, and God could be incarnated in the alpen glow, the mountaineer could incarnate the mountains as when Muir (1987, 15) states, "we are now in the mountains and they are in us" to the point of merging with nature in a romantic, oceanic feeling in which Muir (1980, 88; see also 1992, 212) could claim that "you lose consciousness of your own separate existence: you blend with the landscape, and become part and parcel of nature." Muir first published these words in 1872. W. H. Hudson countered Muir's sublime romanticism about 30 years later by suggesting:

> [W]e are no longer isolated, standing like starry visitors on a mountain top, surveying life from the outside, but are on a level with and part and parcel of it. (Finch and Elder 1990, 308)

Rather than the lone wanderer on the mountain top in Caspar David Friedrich's painting, Rachel Carson took up Hudson's cry that humans are part and parcel of nature (as we will see in a later chapter).

For Muir even visitors to mountainous national parks could partake of a sacramental repetition or re-enactment of sublime incarnation. Muir (1991, xvii; my emphases) announced in his "Preface" to his book on national parks that its aim was "to show forth the beauty, grandeur, and all-embracing usefulness of our wild *mountain forest* reservations and parks with a view to inciting people to come and enjoy them, and get them into their heart, that so at length their preservation and right use might be made sure." Muir makes no attempt to disguise the fact that his book is devoted to 'our' *mountain forest* national parks.

Whilst subscribing to some dominant discourses of national parks Muir also flies in the face of others, such as the national parks as useless or worthless lands (the so-called worthless lands thesis) (see Runte 1979, 28, 49 and 60), and national parks as nationalist parks in which America saw and commemorated itself as what Perry Miller (1956, 209 and 1967, 201) calls "nature's nation." Against both discourses, Muir advocates a kind of spiritual and sacramental incarnation of nature into the national heart or psyche unlike the material incorporation of nature into and by the gaping maw of the national, capitalist state and industrial technology (see MacCannell 1992, 115). Yet as nature's nation also saw itself as God's nation, national parks were God's national parks not only for Muir but also for the nation at large. Yet the view of national parks in the US as "America's secular cathedrals" persists with a previous Secretary of the Interior, Bruce Babbitt (cited in Mitchell 1994, 12), describing them exactly in these terms.

Not only has "[n]ature become a secular deity in this post-romantic age" as William Cronon (1996, 36) puts it, but National Parks are America's secular cathedrals in which this deity is worshipped under the sign of the sublime. Muir's work in particular is caught up, as are national parks in general, in this triangular force field between, even in this American cultural and political triumvirate of, God, nature, and nation with the national park as their point of intersection in what could be called 'God's nation's nature.' The American national park is thus not nature in the raw but nature cooked by culture. Muir repeats what Lawrence Buell (1995, 62) calls "the gesture of putting the nation under the sign of the natural: America as crag, Canada as iceberg, Australia as outback."

For Muir, national parks as nature incarnated, rather than nature incorporated, will not merely be places out there to 'enjoy' but places to which 'people' (though the question is raised of *which* people? as is the question of *whose* national parks? in relation to the title of his book—who does "our" refer to?) belong in a *natural* nationalism and not so much places which belong to the people in *cultural* nationalism. By taking this position Muir sought to reverse the human/nature dualism so endemic to western culture, even subvert the God/nature (including humans) dualism of Calvinism, and replace the democratic ideal with a nature-cratic one ruled by "the parliament of trees" and the law of Nature, rather than the law of God, and supersede cultural nationalism with natural internationalism which would transcend nations and national borders.

Muir also sought to overcome a narrowly utilitarian view of forest reservations, the predecessor of national parks. Against the prevailing view of his time that mountainous landscapes were 'worthless lands' and that forest reservations were only useful for timber Muir (cited by Runte 1979, 62) posed what he called "the use of beauty." Of a mountain wild(er)ness experience in general he wrote:

> [T]he tendency nowadays to wander in wilderness is delightful to see. Thousands of tired, nerve-shaken, over-civilized people are beginning to find out that going to mountains is going home; that wildness is a necessity; and that mountain parks and reservations are useful not only as fountains of timber and irrigating rivers, but as fountains of life. (Muir 1991, 1; see also 217)

Muir equates wilderness with, and reduces it to, mountains. Muir also sees the mountains (and wilderness unlike his colonizing compatriots) as home.

Like one of the other 'grandfathers' of the American conservation movement in Thoreau (Tilden 1970, 303 called Muir "the Western Thoreau"), Muir praised the prophylactic features of wild(er)ness and a wild(er)ness experience. Muir followed the motto of the wilderness wanderer penned by Thoreau (cited by Hoagland 1987, 48): "in wildness is the preservation of the world." Yet Thoreau viewed wilderness more broadly than Muir (1991, 61, 70 and 74) who generally conflates wilderness, mountains, and the sublime with their peaks figured as "colossal spires" of "massive sublimity." This conflation is summed up in Muir's (1991, 70) reference to "a sublime wilderness of mountains" and to "the massy sublimity of the mountains" (Muir 1987, 141). Muir (1980, 124) also refers to "a sublime wilderness of mountains" in his essays. By doing so Muir enacts a masculinist construction of nature and reality in which the glacial, masculine, homely, and sublime of the mountains are seen as the origin of life rather than the littoral, feminine, matrifocal (un)homely, and slime of the swamp or other wetland.[7]

Yellowstone National Park impressed Muir so much that he regarded climbing its mountains as a sublime experience. Climbing Electric Peak for Muir (1991, 44) is a means to "get yourself kindly shaken and shocked. You are sure to be lost in wonder and praise, and every hair of your head will stand up and hum and sing like an enthusiastic congregation." Muir, like Walt Whitman, sings the body electric (or more precisely, it sings him) of the individual, masculinized human body in a patriarchal hymn to God the Father Law, not to 'Mother Nature.' The sublime is an individual, not a communal, experience.

For Thoreau and Muir wild(er)ness, rather than the primitive or the barbarian with their pejorative overtones, was distinct from civilization (Muir 1991, 2). In between the wild(er)ness of mountainous national parks and the civilization of the cities Muir (1991, 2) places "the half wild parks and gardens of towns," though for him national parks are "worth infinitely more than all the gardens and parks of town" (Muir 1979, 351). Muir (1991, 2), however, sees this care for the parks and gardens of town as encouraging. He even finds encouraging "the scenery habit in its most artificial forms, mixed with spectacles, silliness, and kodaks." Yet the scenery habit is not only mixed with spectacles, silliness, and kodaks but also potentially with the sublime and sublimation as both valorize the sense of sight over all other senses. It is perhaps hardly surprising then that Muir would not disparage the scenery habit but would hope it would lead on in similarly sighted fashion to the sublimatory habit.

[7] I discuss the s(ub)lime and the wetland as womb of the world in Giblett (1996).

If mountainous national parks are a mediating category for Muir (1991, 4) so too are human beings who he describes as "half animal, half angel," half slime, half sublime, or s(ub)lime in Zoë Sofoulis' (1988, 12) parenthetical portmanteau. The angelic is strongly associated for Muir (1991, 56) with mountainous wild(er)ness that is "the dwelling-place of the angels," whereas animals are "fellow beings, so seldom regarded in civilization"; plants also for that matter will "soon come to be regarded as brothers" (Muir 1979, 350). Animals are even regarded by him as "brimful of humanity" (Muir 1991, 59) as is "the whole wilderness [which] seems alive and familiar, full of humanity" (Muir 1987, 238). Muir's assessment of plants, animals, and wilderness smacks of anthropocentrism, but it may have been a tactical move to see them at least in human terms and not less than, or alien to, the human.

10.3 Or the Great Goddess's?

Yet Muir also propounded a competing view of nature sacralized to that of his naturalistic Campbellism. He concedes that it is not just mountains and forests that are enjoyable and inspirational, but all wilderness. Muir stated categorically that "none of Nature's landscapes are ugly so long as they are wild" (Muir 1991, 4). As wildness is a matter of point of view, rather than a quality of substance (see Giblett 2011, chapter 5), Muir would seem to subscribe to Thoreau's (1962, *VIII*, 275) dictum that "the same object is ugly or beautiful, according to the angle from which you view it." Indeed, according to the distance from, and the angle at which, we view it as (see Giblett 2011, chapter 3) and our pre-disposition towards it as Thoreau (1962, *IX*, 466) maintained later that "we find only the world we look for." The world we look for is the world we find. We cannot find what we do not see.

Beautiful (because wild) landscapes could include swamps as Muir (1991, 5) discovered in his conversion to conservationism, or at least to sanctuarism: "*Calypso borealis* still hides in the arbor vitae swamps of Canada, and away to the southward there a few unspoiled swamps, big ones, where miasma, snakes, alligators, like guardian angels, defend their treasures and keep them as pure as paradise." Yet rather than separating Muir off as an original figure, this conversion experience in the wilderness makes him fit the pattern of the frontier hero as outlined by Richard Slotkin (1994, 374). According to Wolfe (1945, 146–147), Muir's encounter with the *Calypso borealis* in a Canadian marsh was one of "the two supreme moments of his life," the other being his meeting with Emerson, and not his meetings with Roosevelt, the Sierras or sequoias (as we will see in a later chapter). The swamps referred to as "away to the southward" were presumably the swamps of Florida, Muir's encounter with an alligator in one he devotes one chapter of his *A Thousand-Mile Walk to the Gulf* (as we saw in the previous chapter; see also Worster 2008, 93–95).

Muir propounded not only a sublime aesthetics of the mountains, geysers, and glaciers but also a watery and earthy (slimy) experience of the wetlands. Muir (cited by Runte 1979, 31) looked forward to the day when "lowlands will be loved more than alps, and lakes and level rivers more than waterfalls." That day has far from arrived. Mountains and waterfalls are still a dominant feature of the modern aesthetics of landscape. Yet lowlands for Muir (1987, 186) included not only wetlands but also cities with their "lowland care and dust and din, where Nature is covered and her voice smothered." Perhaps he was implying that cities should also be loved more than alps and waterfalls.

Muir's opposition to the narrowly utilitarian point of view of nature and national parks can also be seen in his answer to the rhetorical question, "what are rattlesnakes good for?" Muir (1991, 25) replies that they are "good for themselves" (as a national park should be looked after "for its own sake" and "we need not begrudge them [the rattlesnakes] their share of life") (Muir 1991, 43). Muir's position on rattlesnakes (and alligators) has been seen by some as the foundation for a biocentric ethic (as we have seen). Muir's anthropocentrism in his aims for his book and in his attitude to plants and animals in general and his biocentrism in relation to rattlesnakes in particular were contradictory positions, but could also be seen as tactical moves depending on the object under discussion and the entrenched attitudes to them to be either extended (in the case of humanism to plants and animals) or reversed (in the case of rattlesnakes).

Yet by converting God into nature and by subscribing to biocentrism rather than to anthropocentrism, Muir was staying within and reproducing their categories. Mountainous, sublimated nature in national parks was still God the Father Law for Muir though in his pantheon he could also allow for a slimy, desublimated unnamed Palaeolithic Great Goddess in the swamps and a good Neolithic 'Mother Earth' or 'Mother Nature' everywhere. For Muir (1980, 229), "Nature is a good mother" as well as a bad father, or more precisely nature for him is split between the two.

Muir's work on, or more precisely for, mountainous and/or forested, national parks marks a shift from an anthropocentric to a biocentric ethic for national park and wild(er)ness areas, though Muir's article on "The American Forests" for *The Atlantic Monthly* (republished in *Our National Parks*) showed, as Roderick Nash (1982, 136) points out, "Muir's continued ambivalence on the 'forestry-or-preservation issue'." Muir's other articles of the 1890s for *The Atlantic Monthly*, particularly "The Wild Parks and Forest Reservations of the West" (also republished in *Our National Parks*; see Muir 1991), are far less ambivalent on this issue and far more committed to the preservation side of the struggle.

Muir saw that the wild(er)ness, and not just mountain wilderness, was already safer than civilization. The implication is that so-called civilization, not wild(er)ness, needs to be made safe. Despite the snakes and alligators and other natural predators of the swamp wild(er)ness, Muir (1991, 21) thought that "it is far safer to wander in God's woods than to travel on black highways or to stay at home. [...] No American wilderness that I know of is so dangerous as a city home 'with all the modern improvements.' One should go to the woods for safety, if for nothing else." Perhaps one should not go to swamp for safety, nor to the woods and mountains, though the latter could provide the terrifying and uplifting experience of the sublime, whereas the former could mean a horrifying and down putting immersion in the slimy.

For Muir, one should go to the woods not only for physical safety but also for spiritual succour. In fact, for him "what we call parks" would better be called "places for rest, inspiration, and prayers" (Muir 1991, 23). Wooded national parks for Muir are sacred places in which, as Rupert Sheldrake (1991, 23) suggests, "the spiritual and the physical are experienced together. Sacred places are openings between the heavens and the earth." Muir certainly saw mountainous and wooded national parks as sacred places in which the spiritual and the physical could be experienced together via a sublime opening between the heavens and the earth. He even saw national parks as sacred places where it would be possible to overcome the dualism between the physical and the spiritual. Indeed, on the one hand, he is largely a Platonist in his reference to "the truly substantial, spiritual world whose forms flesh and wood, rock and water, air and sunshine, only veil and conceal" and, on the other, a naturalistic Campbellite in his preference for the immanence, or even incarnation, of the spiritual in the physical for "here [in Yellowstone National Park] is heaven and the dwelling-place of angels [...] the terrestrial manifestations of God" (Muir 1991, 55–56).

Muir seems to have sensed no contradiction between these two positions expressed together within a page of each other because for him there was no contradiction between them. Muir saw an exchange and mutuality between the spiritual and the material exemplified in the forested and/or mountainous national park. He refused the views of the capitalists, scientists, engineers, and technocrats of his time who saw nature as so much raw material and dead matter to be manipulated and exploited, and of the Christians who saw nature as so much clinging and sinful materiality to be sloughed off and transcended. For him nature was living, divine, dynamic yet it also had its stern and craggy, patriarchal and Campbellite God up in the mountains and its wise, tender, though also stern (see Muir 1987, 142) and matrifocal Great Goddess down in the swamps. Nature for Muir was bifurcated, even trifurcated with the third figure of a good Mother Earth or Mother Nature.

Muir subscribed not just to "the immanence of nature in the divine" as Sheldrake (1991, 197) puts it, but also to the transcendence of the divine in nature. Muir was thus, in Sheldrake's (1991, 198) terms, a panentheist for whom "God is not remote and separate from nature, but immanent in it. Yet at the same time, God is the unity that transcends it. In other words, God is not just immanent in nature, as in pantheist philosophies, and not just transcendent, as in deist philosophies, but both immanent and transcendent, a philosophy known as *panentheism*." Or should that be Goddess, rather than God? Zimmerman (1994, 254) maintains that "goddess spirituality holds that the divine is both immanent [...] and transcendent." The divine for Muir overarches nature, transcends everything, but is immanent in individual, particular places, and not just in forested and/or mountainous national parks but also in swamps; the divine is cosmically and locally transcendent and immanent; the whole cosmos and every locality manifests the divine, the divine transcends the local, the divine is in the local, and vice versa. Muir was no mere radical who "thought globally and acted locally," but a panentheist who sought and thought the divine in the local and enacted the local in the divine.

10.4 Writing the Word of Nature

If national parks function as modern cathedrals with their "sermons in stones" (Muir 1991, 59), then their sacred scriptures are the inscriptions of nature *in* the earth (unlike the writing of the modern, colonial city *on* the earth, and unlike the slimy tracery of the Great Goddess, the swamps, the womb of the earth) (see Giblett 1996, especially chapter 4). Muir (1991, 45) advises his readers that "after this reviving experience [of climbing Electric Peak], you should take a look into a few of the tertiary volumes of the grand geological library of the park [Yellowstone National Park], and see how God writes history." Muir was the culmination of a nineteenth-century American phenomenon in which, as Perry Miller (1956, 211 and 1967, 203) puts it, "nature somehow [...] had effectually taken the place of the Bible." Or more precisely, as nature had taken the place of God for Muir, so the Word of Nature had taken the place of the Word of God. Muir was a self-mythologized literary frontier hero who was, in Richard Slotkin's (1994, 374) terms, "generally disinclined to learn from book culture when the book of nature is free to read before him." Printing made the word of God available to the vernacular reader whereas Muir found the book of nature already open ready for him to read freely.

In the modern mountainous wild(er)ness one can see not how God writes the law in words on tablets of stone as "He" did in the Judaic wilderness with Moses, but how "God" writes history in "words" of stone in "the divine manuscript" (Muir 1987, 132) of mountainous nature as "He" does in the law of Muir, the modern-day Moses of mountainous national parks. For Muir (1991, 45), here in Yellowstone National Park are "a wonderful set of volumes lying on their sides,—books a million years old, well bound, miles in size, with full-page illustrations [...] telling wonderful tales of bygone centuries." For Muir (1991, 59), the "natural history," or better national park, or even better still mountainous wild(er) ness writer, "to get all this ['the wonderful clearness and freshness of the rocky pages'] into words is a hopeless task." As a result, "to defrauded town toilers, parks in magazine articles are like pictures of bread to the hungry. I can write only hints to incite good wanderers to come to the feast" (Muir 1991, 59). Rather than the oral sadistic feast of capitalism, the feast Muir had in mind here was more a passover feast, a sacramental communion. Muir as a magazine article writer was the frontier hero of a literary mythology through whom, as Slotkin (1994, 374) puts it, "the deep wisdom of the natural wilderness is transmitted to the sluggish life of the Metropolis."

One course of the feast, to switch metaphors, would be a richly layered text. In these books of nature "the post-glacial agents [of air, rain, frost, rivers, earthquakes, avalanches (Muir 1991, 63)] are at work on the grand old palimpsest of the park region, inscribing new characters" (Muir 1991, 49) on the *tabula rasa* or "white bosom" of the glaciers rather than tracing stories in the living black waters of wetlands. The book of nature is not set in stone for all time but is constantly changing with new writings being traced over old ones to produce a deeply and richly patterned palimpsest. Muir's own writing on glaciers is one layer of the palimpsest inscribed on its broad white bosom. Glaciers themselves may deserve "the name of Destroyer [...] but we quickly learn that destruction is creation" (Muir 1991, 72) and "we see that everything in Nature called destruction must be creation" (Muir 1987, 229). That natural (not human) destruction is creation is one of Muir's favourite maxims (see Muir 1987, 238 and 1980, 31 and 53).

Not only glaciers and post-glacial agents write the text of the book of nature, but all inhabitants of the wild(er)ness do, even the swarming insects are "a cloud of witnesses telling Nature's joy" and the plants are "singing the old new song of creation." All creatures have a kind of evangelistic function of spreading the gospel according to Muir (1991, 52–53) about the religion of nature. Indeed, "each and all tell the orderly love-beats of Nature's heart," "the heart of nature, whence we came" (Muir 1980, 46), couched within "the generous bosom of the woods" (Muir 1992, 155–156), nature's good breast rather than nature's bad breast of the black waters of wetlands.

Similarly, for Muir (1991, 57) in terms reminiscent of Gerard Manley Hopkins, the Sierra Nevada mountains in Yosemite National Park are "pervaded with divine light, every landscape glows like a countenance hallowed in eternal repose; and everyone of its creatures [...] is throbbing and pulsing with the heartbeats of God." This creative work of nature is always going on with destructive work: "Nature is ever at work building and pulling down, creating and destroying, keeping everything whirling and flowing, allowing no rest but in rhythmical motion, chasing everything in endless song out of one beautiful form into another" (Muir 1991, 73). Nature here is always at work producing not so much a cosmos as a cosmogenesis. Muir's work here is remarkably prescient, even prophetic, of the paradigm shift from modern nature as (dead) machine to postmodern nature as living community, even communion (see, e.g., Swimme and Berry 1992).

Nature is a homely living agent for Muir rather than unhomely dead matter or (un)homely Great Goddess. Going to the woods for Muir (1991, 74) is going home, or "going to the mountains is like going home" (Muir 1980, 118); going to a national park (but not going to a swamp) for him is "getting in touch with the nerves of Mother Earth" (Muir 1991, 2). Nerves communicate with the whole body through electrical impulses, so getting in touch with the nerves of the body electric of Mother Earth meant getting in touch with the fibres that communicate with the whole body of Mother Earth, both inside and outside the national park.

Yet getting in touch with the nerves of Mother Earth meant *not* getting in touch with her whole body. Going *into* the swamp would be getting in touch with the body of Mother Earth. Perhaps it was in going to the mountains, rather than to the national park per se, that Muir got in touch with the sublimated body of Father Law. The mountains could be incarnated in Muir, and Muir in the mountains, so that there was no disconnection between the two and he became "an inseparable part of it [...], a part of all nature" (Muir 1987, 16). In "the Sierra Cathedral," for Muir (1987, 243), "we feel ourselves part of wild Nature, kin to everything."

If Mother Earth for Muir has nerves she also has a heart, what Muir (1991, 4) calls "the warm, unspoilable heart of the earth," "Nature's warm heart" (Muir 1991, 180). For Muir, the warm, good breast of "Mother Earth is ever familiar and the same" (Muir 1991, 37), unlike the hot workshops of geysers where Father God Law is ever unfamiliar and different, and unlike the hot netherlands of swamps where the Great Goddess is ever uncanny and (un)homely. "Nature's sources never fail" (Muir 1991, 42). The good breast of what Muir calls in a letter the "bosom" of "sweet kindly Mother Earth" (Badè 1924, *II*, 267) never dries up and the (glacial and/or wetland?) womb generously gives birth demonstrating "the infinite lavishness and fertility of Nature—inexhaustible abundance" (Muir 1987, 242). Nature, for Muir (1980, 229), and in short, is "a good mother," though also a bad Father (like his own father about whom we hear so much in the biographies, whilst Muir's "Mother Earth" is perhaps like his mother about whom his readers hear so little, except for his surrogate mother in Jeanne Carr). This lack of "a parental tie to anchor him to the Metropolis," or to metropolitan culture more generally, helps to make Muir fit Slotkin's (1994, 374) model of the frontier hero. Muir (1991, 268) prefers the warm, living nature of Mother Earth, a nurse maid nature "always ready to heal every scar" "one touch" of whose hand "makes the whole world kin" (Muir 1991, 164), to "the grey, savage wilderness of crags and peaks," the visage of his mountainous God the Father Law, who "seems lifeless and bare" (Muir 1991, 125).

If Mother Earth has nerves and a heart, then Father Mountain has a face or more precisely a "divine landscape-countenance" with "every feature glowing" (Muir 1987, 15 and 115). For Muir, Father God mountain is the body of nature from the neck up, the site of sublimation, intellection and theorization, the heights of the sublime spires, whereas the Great Goddess and Mother Earth are the body of nature from the neck down, the womb of new life, the organs of nurturing, the depths of the slimy swamps. Muir (1991, 55) remarks on how "the physiognomy and even the complexion of the landscape should still be so divinely fine!" and "all the landscape is glowing like a benevolent countenance at rest" (Muir 1991, 193) or "the landscape [is] beaming with consciousness like the face of a god" (Muir 1987, 85) when referring to the sublimated alpen glow. Muir looks on this face in mountainous and forested national park, the stern, craggy face of God the Father Law and not on the body of the Great Goddess, life-giving and death-dealing. Political ecocriticism of his nature writing brings these aspects into the open of psychopolitical ecology and awakens the reader to them.

References

Badè, William. 1924. *The Life and Letters of John Muir, Volumes I and II*. Boston: Houghton Mifflin.

Benjamin, Walter. 1986. *Moscow Diary*, ed. Gary Smith. Translated by R. Sieburth. Cambridge, MA: Harvard University Press.

Bookchin, Murray. 1991. *The Ecology of Freedom: The Emergence and Dissolution of Hierarchy*, revised edition. Montreal: Black Rose.

Buell, Lawrence. 1995. *The Environmental Imagination: Thoreau, Nature Writing, and the Formation of American Culture*. Cambridge, MA: Harvard University Press.

Butcher, Devereux. 1969. *Exploring Our National Parks and Monuments*, 6th ed., Revised. Boston: Houghton Mifflin.

Cohen, Michael. 1984. *The Pathless Way: John Muir and American Wilderness*. Madison: University of Wisconsin Press.

Cronon, William. 1996. Introduction: In Search of Nature. In *Uncommon Ground: Rethinking the Human Place in Nature*, ed. William Cronon, 23–66. New York: W. W. Norton.

Day, David. 1996. *Claiming a Continent: A History of Australia*. Pymble, NSW: Angus and Robertson.

Eagleton, Terry. 1990. *The Ideology of the Aesthetic*. Oxford: Basil Blackwell.

Finch, Robert, and John Elder, eds. 1990. *The Norton Book of Nature Writing*. New York: W. W. Norton.

Flichy, Patrice. 1995. *The Dynamics of Modern Communication: The Shaping and Impact of New Communication Technologies*. Translated by L. Libbrecht. London: Sage.

Fox, Stephen. 1981. *John Muir and His Legacy: The American Conservation Movement*. Boston: Little, Brown.

Giblett, Rod. 1996. *Postmodern Wetlands: Culture History Ecology*. Edinburgh: Edinburgh University Press.

Giblett, Rod. 2008. *Sublime Communication Technologies*. Basingstoke: Palgrave Macmillan.

Giblett, Rod. 2011. *People and Places of Nature and Culture*. Bristol: Intellect Books.

Giblett, Rod. 2018. *Environmental Humanities and Theologies: Ecoculture, Literature and the Bible*. London: Routledge.

Gilpin, William. 1794. *Three Essays: on Picturesque Beauty; on Picturesque Travel; and on Sketching Landscape […]*. 2nd ed. London: A. Blamire.

Grove, Richard. 1995. *Green Imperialism: Colonial Expansion, Tropical Island Edens and the Origins of Environmentalism*. Cambridge: Cambridge University Press.

Hoagland, Edward. 1987. In Praise of John Muir. In *On Nature: Nature, Landscape, and Natural History*, ed. Daniel Halpern, 45–58. San Francisco: North Point Press.

Lane, Belden. 1988. *Landscapes of the Sacred: Geography and Narrative in American Spirituality*. Mahwah, NJ: Paulist Press.

MacCannell, Dean. 1992. Nature Incorporated. In *Empty Meeting Grounds: The Tourist Papers*, 114–117. London: Routledge.

MacKenzie, John. 1988. *The Empire of Nature: Hunting, Conservation and British Imperialism*. Manchester: Manchester University Press.

Miller, Perry. 1956. Nature and the National Ego. In *Errand into the Wilderness*, 204–216. Cambridge, MA: Belknap Press of Harvard University Press.

Miller, Perry. 1967. The Romantic Dilemma in American Nationalism and the Concept of Nature. In *Nature's Nation*, 197–207. Cambridge, MA: Belknap Press of Harvard University Press.

Mitchell, John. 1994. Our National Parks. *National Geographic* 186 (4): 2–55.
Muir, John. 1979. Thoughts upon National Parks. In *John of the Mountains: The Unpublished Journals of John Muir*, ed. Linnie Wolfe, 350–354. Madison: University of Wisconsin Press.
Muir, John. 1980. *Wilderness Essays*, ed. Frank Buske. Salt Lake City: Peregrine Smith.
Muir, John. 1987. *My First Summer in the Sierras.* New York: Penguin. First published 1911.
Muir, John. 1991. *Our National Parks.* San Francisco: Sierra Club Books. First published 1901.
Muir, John. 1992. *A Thousand-Mile Walk to the Gulf.* New York: Penguin. First published 1916.
Nash, Roderick. 1982. *Wilderness and the American Mind.* 3rd ed. New Haven: Yale University Press.
Neumann, Roderick. 1995. Ways of Seeing Africa: Colonial Recasting of African Society and Landscape in Serengeti National Park. *Ecumene* 2:149–169.
Runte, Alfred. 1979. *National Parks: The American Experience.* Lincoln: University of Nebraska Press.
Schama, Simon. 1995. *Landscape and Memory.* London: HarperCollins.
Sears, John. 1989. *Sacred Places: American Tourist Attractions in the Nineteenth Century.* New York: Oxford University Press.
Sheldrake, Rupert. 1991. *The Rebirth of Nature: The Greening of Science and God.* New York: Bantam.
Slotkin, Richard. 1994. *The Fatal Environment: The Myth of the Frontier in the Age of Industrialization 1800–1890.* New York: HarperCollins.
Sofoulis, Zoë. 1988. *Through the Lumen: Frankenstein and the Optics of Re-Origination.* PhD Diss., University of California, Santa Cruz.
Swimme, Brian, and Thomas Berry. 1992. *The Universe Story: From the Primordial Flaring Forth to the Ecozoic Era – A Celebration of the Unfolding of the Cosmos.* San Francisco: Harper.
Thomas, Keith. 1984. *Man and the Natural World: Changing Attitudes in England 1500–1800.* Harmondsworth: Penguin.
Thoreau, Henry David. 1962. In *The Journal 1837–1861, Volumes I–XIV*, ed. Bradford Torrey and Francis Allen. New York: Dover.
Thoreau, Henry David. 1980. *Natural History Essays.* Salt Lake City: Gibbs Smith.
Tilden, Freeman. 1970. *The National Parks*, revised and enlarged edition. New York: Alfred A. Knopf.
Turner, Frederick. 1985. *Rediscovering America: John Muir in His Time and Ours.* New York: Viking.
Virilio, Paul. 1989. *War and Cinema: The Logistics of Perception.* Translated by P. Camiller. London: Verso.

Wolfe, Linnie. 1945. *Son of the Wilderness*. New York: Alfred A. Knopf.

Worster, Donald. 1993. *The Wealth of Nature: Environmental History and the Ecological Imagination*. New York: Oxford University Press.

Worster, Donald. 2008. *A Passion for Nature: The Life of John Muir*. Oxford: Oxford University Press.

Zimmerman, Michael. 1994. *Contesting Earth's Future: Radical Ecology and Postmodernity*. Berkeley: University of California Press.

CHAPTER 11

Nature Writing and the Politics of Ecocides and Environmental Conservation: Rachel Carson

Rachel Carson was born in 1907 and died in 1964 tragically of cancer. She was a marine biologist by training and the mother of American conservation by writing. Henry David Thoreau can be considered the great-grandfather of American conservation, John Muir the grandfather, and Aldo Leopold the father. This lineage and these roles do not imply any sort of close or intimate relationship between Leopold and Carson whose lives were separated by more than a decade and the divide between the western, mid-western, and eastern US with Leopold spending most of his life in the mid-west and some of it in the west, and Carson on the eastern seaboard. He was born 20 years before her and died 16 years before her. Both lived too short a life; both were proponents of conservation in the broader sense of conserving not only special species and places, but all life on planet earth.

As a marine biologist by training and profession of the intertidal zone where land and water meet and mingle on the edge of seas and oceans, Carson is the unlikely patron saint of American conservation. According patron sainthood to her is not a matter of hagiography, of producing an idealized portrayal of her that glosses over her blind-spots and ignores her feet of clay as a saint. As a marine ecologist of the intertidal oceanic zone, but not of liminal and littoral marshes, she had a blind-spot about wetlands. Her canonization rests on her classic environmental book *Silent Spring* first published in 1962 in which she presented a well-documented

R. Giblett, *The Politics of Nature and Environmental Writing*, Transdisciplinary Environmental Humanities 1,
https://doi.org/10.1007/978-3-032-04427-3_11

indictment of the use of the insecticide DDT and other biocides in the US and voiced a powerful protest against DDT and the others, and their uses. *Silent Spring* is usually read today as an anti-DDT tract and as a green consciousness-raiser. It is often seen as the founding text of the modern American environmental movement and as being broadly influential in the latter half of the twentieth century. Former Vice-President and Presidential candidate Al Gore (1994, xxv) in his foreword to a reissue of Carson's classic relates how, in 1992 on the thirtieth anniversary of the publication of *Silent Spring*, "a panel of distinguished Americans selected *Silent Spring* as the most influential book of the last fifty years."

11.1 Carson's Ecological Critique of Masculinism and Militarism

Like Henry David Thoreau, who was more than just a surveyor of snow storms, a contributor to his own journal with a circulation of one, a leader of a huckleberry-gathering excursion, and a lover of swamps and their patron saint, and Aldo Leopold, who was more than just an expert ecologist as a wildlife manager and lover of marshes and their patron saint, Carson was more than just a nature lover of the intertidal zone and accredited expert ecologist as a marine biologist (Carson 1998, especially 2–4). If she had remained a marine biologist and a nature writer, she would not have become the patron saint of American conservation. She would be remembered and canonized as a nature writer (see Finch and Elder 1990, 519–524), and not also as an environmental writer (see McKibben 2008, 365–376), nor as what Felix Guattari (1989, 140; 2000, 52) calls a third ecologist: "ecology must stop being associated with the image of a small nature-loving minority or with qualified specialists. Ecology in my sense questions the whole of subjectivity and capitalistic power formations," including the militarization of all life and the imposition of the corporate-state cult of death. Feminist political ecology follows suit with the added emphasis on gender that the militarization of all life and the imposition of the corporate-state cult of death are masculinist. Carson's ecology with its critique of masculinism and militarism contributed these vital facets to the creation of feminist political ecology that embraces nature and gender critically.

Perhaps more than Carson's exposé of chemical abuse, her critique of the militarization of all life and the imposition of the corporate-state cult of death are what really got up the noses of the chemical corporations. The main reason that Carson attracted and endured the ire of corporate bosses and their governmental allies was not just, or even because, she criticized the use of, and the damage done by, insecticides such as DDT as many had done that before her (and she was drawing and relying on already published sources), but because she critiqued the militarization of post-World War II, Cold War American society, and its death-drive in a clearly written and cogently argued manner written in a conversational and impassioned style accessible to most readers in a book that became a best seller and was put on the high school English syllabus. Each of these points was held against her and *Silent Spring*. Carson was the critic of a kind of Pearl Harbour-style blitz against nature including plants, animals, and human beings that was delivered from the sky by excess planes left over after World War II (1939–1945) had finished. Modern industrial chemistry for Carson (1994, 7) in *Silent Spring* is war against nature with a corporate-state cult of death inflicted on all living beings, including humans.

Carson devotes her third chapter of 20 pages in *Silent Spring* to "Elixirs of Death" and discusses various death-dealing industrial chemicals. An elixir is a magical or medicinal potion that has its roots in alchemy. Modern industrial chemists eschewed their lineage with alchemy, magic, and potions. They were concocting medicinal compounds to cure the earth of 'pests,' not indulge in fantasies of turning base metals into gold, though they were engaged in the industrial transformation of chemical elements into money. Even more alarming for the modern industrial chemist, their corporate bosses, and government allies is that Carson in the conclusion to the trenchant critique in the chapter "And No Birds Sings" pointed the finger of blame, not at them in particular, for unleashing an "ever-widening wave of death," but at "the authoritarian" with unbridled power (Carson 1994, 127). Carson presciently sensed the presence of this figure whose power and influence was present in World War II living on in the Cold War, and today. This "wave of death" flows into "rivers of death," the title of the following chapter of *Silent Spring* with "dead and dying fish" in "a picture of death and destruction" (Carson 1994, 131). A cult of death indeed. Rivers and wetlands of living water have become streams and habitats of death.

In *Silent Spring* Carson (1994, 3) was prepared to confront, and convey, that "a grim spectre has crept upon us almost unnoticed." This grim spectre, whether it be merely the Grim Reaper of death, or a state or corporate cult of death, or both, was the ghost who haunted the battlefields of what Carson (1994, 7) called "man's [*sic*] war against nature." Not only the body, but also the earth had become a battlefield. The battlefields of the earth were no longer confined or limited to the fields of war. The earth was not only the ground on which the war against nature took place but also the home of the enemy—nature. The war against nature included for Carson (1994, 34) "our war against insects" and "our simultaneous war against the weeds." It was also a war against "man" himself for "man, however much he may like to pretend to the contrary, is part of nature" (Carson 1994, 188). If, as Carson (1994, 86) puts it, "man's [*sic*] announced goal" is "the conquest of nature," then he must also aim to conquer himself in a self-defeating assault. When he mounts what Carson (1994, 74) calls a "chemical assault" against nature, he mounts a chemical assault against himself for he is a chemical compound, a biogeochemical being.

This chemical assault on nature is recent, a bastard child of industrial capitalism coupled with its technologies. Carson relates that:

> with the dawn of the industrial era the world became a place of continuous, ever-accelerating change, a smooth superhighway on which we progress or proceed with great speed. Instead of the natural environment there was rapidly substituted an artificial one composed of new chemical and physical agents, many of them possessing powerful capacities for inducing biological change. (Carson 1994, 219–220 and 277)

Many of these capacities were unknown, or were only known in relation to a specific use or application, and rarely in a broader context, or in combination with other chemical substances. Carson (1994, 187) relates that "as the tide of chemicals born of the Industrial Age has arisen to engulf our environment, a drastic change has come about in the nature of the most serious public health problems. [...] Today we are concerned with [...] a hazard we ourselves have introduced into our world as our modern way of life has evolved." Modern industrial technology has produced chemicals that are hazardous, not only to the rest of the world, but also to humans. "Man," for Carson (1994, 219; her emphasis) "alone of all forms of life, can *create* cancer-producing substances, which in medical terminology are called carcinogens." Pesticides turn not only the landscape into a battlefield but also the human body. What Carson (1994, 219, 221 and 242) calls "the battle of living things against cancer" began a long time ago but it reached new heights with "that malignant disease [threatening to] strike" "one in every four" individuals (now one or two in three) and "two out of three families" (now more).

Hypermodern medicine has waded, and weighed, into this metaphorical landscape of the body individual and politic to wage war against disease. Hypermodern medicine fights what Carson (1994, 240 and 266) calls "battles in the war against disease" hoping eventually to win "the triumphant war against disease." Hypermodern medicine is militaristic. Post-World War II medicine is part of the militarization of civilian life. Post- World War II natural 'resource management' is also part of the same militarization of civilian life. Carson (1994, 296) couches the alternatives in the same terms as for her "there is a whole battery of armaments available to the forester who is willing to look for permanent solutions that preserve and strengthen the natural relations in the forest." In seeking to critique the militarism of pesticide spraying she uses the same rhetoric in advancing alternatives and is unwittingly complicit with the militarization of language.

Not only 'man' himself but also others are caught up in what Carson (1994, 94) calls "the chemical war." This war is unwinnable not only because the enemy develops resistance and fights back but also because friends are killed outright. For Carson (1994, 8) "the chemical war is never won, and all life is caught in its violent crossfire." All life suffers 'collateral damage' from 'friendly fire.' Carson argues that (1994, 251) "we have turned our artillery against our friends." For Carson (1994, 99) "the question is whether any civilization can wage relentless war on life without destroying itself, and without losing the right to be called civilized." By doing so we have become more barbarian than the barbarians. The barbarians are not at the gates, but already within the walls of the citadel.

We have turned our artillery against our insect allies who "keep at bay" what Carson (1994, 251 and 246) luridly calls "a dark tide of enemies," a veritable "horde" of insect enemies. Carson employs the metaphor, and raises the spectre, of the barbarian hordes of insects overrunning the ramparts of civilization and trying to take over the world just as they do in *Starship Troopers*. The war against nature, and the chemical war against insects and weeds, was born out of war. The industry for the production of what Carson (1994, 16) calls "man-made or synthetic chemicals with insecticidal properties" is "a child of the Second World War." The insecticide industry is a Bachelor Birth from a Bachelor Machine (see Giblett 2008, 95–98). The child went off to fight the war. Modern insecticides for Carson (1994, 18) are "agents of death."

11.2 The War Against Nature

The war against nature, and the chemical war against insects and weeds, was not only borne of war but is also fought like war using the same technologies, such as planes in aerial spraying. The war against nature was not only fought by military means but also couched in natural terms. Carson likens aerial spraying to rain, albeit synthetic and deadly. Carson (1994, 12) refers repeatedly to "the chemical death rain," to "a rain of chemicals" (1994, 89), "the death rain of poison" (1994, 258) and "a rain of poison falling from the skies into the world of wildlife" (1994, 87), and tame life too. This rain was not the black rain of pollutants pumped into the air that later fell back to earth to poison plants and animals, but the toxic rain of poisons sprayed from planes directly onto plants and animals, including humans.

Carson (1994, 89) relates how "one of the first large-scale attacks on the Japanese beetle from the air" occurred in 1959 in Michigan. Planes were authorized to fly low in a kind of second Pearl Harbour, not in revenge for the first as that had already been inflicted terribly at Hiroshima and Nagasaki, but ironically against the Japanese beetle and the houses and other places that harboured it with a twisted sense of poetic justice. In chapter 10 of *Silent Spring* entitled "Indiscriminately from the Skies" Carson (1994, 155–172) describes how "the abundance of surplus planes after the Second World War" were used for aerial spraying. These demobilized planes were remobilized in the war against nature in which they "showered down" what a British ecologist (cited by Carson 1994, 156) called "'an amazing rain of death' upon the surface of the earth." This rain of poison accumulates and forms what Carson (1994, 146) calls "toxic river [and wetland] water" with "its death-dealing power." If life-giving rain is poisoned, if poisons are rained, and if rivers and wetlands run with poison, then we are engaged in what Carson (1994, 34) called "the poisoning of the environment." In what she calls "the age of poisons" the average consumer can "buy substances of greater death-dealing power" "Beyond the Dreams of the Borgias," the title of chapter 11 of *Silent Spring* (Carson 1994, 174). The average consumer can easily be a greater poisoner than the legendary, self-destroying Italian Renaissance family of the Borgias.

Deliberately drawing on the language of nuclear war and atomic weaponry Carson (1994, 156) argues that "not only the target insect or plant, but anything—human or nonhuman—within range of the chemical fallout may know the sinister touch of the poison." The grim reaper was not merely a spectre of death, but also of death by chemical fallout. Chem War was Cold War by other means. Unlike the Cold War in which nuclear weapons were *not* used but were tested, in the Chem War chemical weapons were used. Chemicals for Carson (1994, 211) "play a role similar to radiation."

Carson also draws on the language of nuclear war and atomic weaponry to describe the outcome of radiation by employing the term "contamination." She critiques the "contamination" of "our world" and she is critical of how our "public waters" are contaminated (Carson 1994, 173 (2x), 176, 179, 181 (3x), 182, 188). To contaminate is to "make impure by contact or mixture; pollute, corrupt, infect." Contamination is "the action of making impure or polluting; defilement, infection." Contamination is a ritual, religious, and health phenomenon with resonances in all three domains. By extension, it can refer to radioactive contamination, linguistic contamination, and textual contamination. By implication it suggests bodily, spiritual, and ecological freedom from pollution, corruption, and infection.

In the post- World War II environmental "holocaust," as Carson (1994, 257) puts it, genocide, or more precisely speciocide, was the war plan deployed against targeted species, such as the gypsy moth. This moth gave its name to a plane only to be rewarded with what Carson (1994, 157) calls the "blanket spraying" (like blanket bombing) of its (and others') habitat with the announced aim of "eradicating" it. Eradication, Carson (1994, 157) points out, "means the complete and final extinction or extermination of a species throughout its range."

In what Carson (1994, 158) calls this "all-out chemical war" against the gypsy moth in 1956 planes "showered down the prescribed DDT in fuel-oil" over nearly a million acres in four states, including such areas as dairy farms, fish ponds, salt marshes, and suburbia "drenching a housewife making a desperate effort to cover her garden before the roaring plane reached her, and showering insecticide over children at play and commuters at railway stations." This scenario is a replay of the iconography of Pearl Harbour—only the arsenal and target is different. Civilians were targeted as in war and "citizens had to suffer the prescribed drenching with DDT" (Carson 1994, 159) just as sheep are drenched with chemicals and just as they suffered nuclear fallout in western states. The US state apparatus regarded its own civilian citizens as a target for a military, Pearl

Harbour-style operation and as docile sheep to be drenched in their techno-pastoral suburbs (see Giblett 2008, chapter 3).

Carson saw broadcast spraying as a crude weapon, like blanket bombing, that killed every living thing in its field and range. Carson (1994, 64) bemoaned "our unthinking bludgeoning of the landscape." She bewailed "the unselective bludgeon of insecticidal poisons" (Carson 1994, 127). For her "as crude a weapon as the cave man's club, the chemical barrage has been hurled against the fabric of life" (Carson 1994, 297). Not content just to wage war against nature by bombing it with insecticides, we bludgeon it into submission, and even blast it into smithereens to destroy the basic building blocks of matter that seem to defy us. We are like the proto-Fascists depicted by Ernst von Salomon who try to destroy matter, to making nothing out of everything, a reversal of God who created everything out of nothing (see Giblett 2008, 93–94).

The scientist bludgeoning of nature is for Carson the cave man of the stone age of science. His mantra of the "control of nature" is for Carson (1994, 297):

> a phrase conceived in arrogance, born of the Neanderthal age of biology and philosophy, when it was supposed that nature exists for the convenience of man [*sic*]. The concepts and practices of applied entomology for the most part date from that Stone Age of science. It is our alarming misfortune that so primitive a science has armed itself with the most modern and terrible weapons, and that in turning them against the insects it has also turned them against the earth.

And against 'man' himself, for 'man' is a part and parcel of nature as W H. Hudson wrote 60 years before Carson.

By taking up arms against 'nature' we take what Carson (1994, 67) calls "the shotgun approach to nature," not only shooting broadcast at all and sundry (including ourselves), but also shooting first and asking questions later. Nature is presumed guilty until proven innocent. The shotgun approach to nature is not even a shotgun wedding, but a shotgun funeral. This approach gives what Carson (1994, 68–69) "a giddy sense of power over nature." According to Carson (1994, 85), "the man with a spray gun" has replaced the man with the six-shooter of the wild west as the culture hero of the war against nature, of the conquest of the wilderness and its indigenous inhabitants. The hero of Chem War against nature is no longer Daniel Boone or Davy Crockett, the heroes of the war against wilderness, but the Wyatt Earp of the spray gun, the Red Baron of DDT, the barnstormer of pesticides. With her critique of masculinism and militarism, Carson's ecology is political and an early contributor of these vital facets to its creation.

As spraying for Carson (1994, 135) is "a weapon," the sprayed landscape is a battlefield after the battle. Following spraying and bird deaths in Norfolk in the UK in 1960 a landowner wrote that "the place is like a battlefield" (cited by Carson 1994, 123). Carson (1994, 71) critiques "the sterile and hideous world we are letting our technicians make." But more than aesthetic considerations are at stake, as Carson (1994, 72) argues, in the preservation of roadside vegetation. For her "in the economy of nature the natural vegetation has its essential place" (Carson 1994, 72). Economy of nature, or ecology, takes precedence over the aesthetics of nature (unlike the Romantics for whom aesthetics took precedence over the economy of nature; see Giblett 2011, chapter 3).

11.3 Carson's Ecology of Bodies and Waking to Nature

For humans and other animals and for Carson (1994, 189) the effects of pesticides on species other than the target one "is a problem of ecology, of interrelationships, of interdependence [...], [of] the web of life—or death," of life and death. There is an ecology of plants and animals, soils, air and water, but there is also what Carson (1994, 189) calls "an ecology of the world within our bodies." For her there is also an aesthetics and a mechanism of the world within our bodies. She refers to "the extraordinary energy-producing mechanism of the body" and to "its beautifully functioning mechanism" (Carson 1994, p 200). Aesthetics re-asserts itself in constituting the body as beautiful (small, smoothly functioning, organic, or organicist) object.

Modern western medicine also asserts itself in constituting the body as (energy-producing) mechanism. What Carson (1994, 201) calls "the transformation of matter into energy in the cell is an ever-flowing process" necessary for life, but it is not necessarily a beautiful, mechanistic process. This is to figure life in aesthetic and mechanistic metaphors. "The balance of nature" for Carson (1994, 246) is "a complex, precise, and highly integrated system of relationships between living things [...] in a constant state of adjustment." Nature for Carson is, in other words, a machine. The balance of nature is homeostatic and cybernetic. Ecology for her is earthly mechanics. Earthly mechanics takes precedence over landscape aesthetics. Life, of course, can only be figured in metaphor, but there is a world of difference between aesthetic and mechanistic metaphors and the Taoist metaphor of life as an ever-flowing process and as a body of energy (see Giblett 2008, chapter 10).

Rather than bludgeoning the landscape, we should, as Carson (1994, 69) puts it, promote "the long-range health of the landscape." For Carson (1994, 78) "soil and the living things in and upon it exist in a relation of interdependence and mutual benefit." Symbiosis and mutual aid, rather than the parasitism and survival of the fittest, are, or should be, the norm (see Giblett 2011, chapter 12). Cooperation rather than competition is the motto of ecological sustainability. The long-range health of the landscape for Carson (1994, 79) depends upon "the health of the soil." Healthy bodies also depend upon healthy land, as do healthy minds (see Giblett 2009, chapter 8).

Silent Spring was first published in 1962, the same year coincidentally in which both William Melvin Kelley's novel *A Different Drummer* and his op-ed article, "If You're Woke, You Dig It" (Gallagher 2021), were published. This year was momentous as it also marked the centenary of Henry David Thoreau's death. These publications of 1962 are a fitting memorial to Thoreau and the legacy of his work. Taken together, they make a joint call for waking up and staying woke. In Kelley's case this meant waking up to racism and social injustice. In Carson's case this meant waking up to what she calls the "now universal contamination of the environment" by chemicals (Carson 2018, 13). In a speech that she gave after the publication of *Silent Spring* she pointed to the hazards of "biocides rather than pesticides" in land, water, and air endangering all life on the earth (Carson 2018, 441, 485). In another speech she described how the industry response to *Silent Spring* of "tranquilizing information [was] designed to lull the public into the sleep from which *Silent Spring* so rudely awakened it" (Carson 2018, 431–432). In a newspaper article Carson (2018, 459) referred to "the now awakened public." Carson with her words of wisdom woke up those of her fellow Americans who were not awake to loving nature. For these and other reasons she is the patron saint of American conservation.

Her work lives on and continues to be influential as Sandra Steingraber indicates in her recent introduction to the Library of America edition of *Silent Spring and Other Writings on the Environment*. Steingraber (2018, xxxiv–xxxv) points to environmental health in general and breast cancer in particular as two areas where *Silent Spring*'s influence is still evident. Other areas where *Silent Spring* is newly influential are "the developments in the fields of agroecology" and "the food sovereignty movement" (Steingraber 2018, xxxix, xlii–xliii n25). Both of these new growths were taking place in the year, and the year before, in which Steingraber was writing as indicated in the books she cites. The words of wise women live on and are a well-spring from which those who come after continue to draw edification and inspiration. Carson is no exception. *Silent Spring* is such a well-spring of well-wishes.

References

Carson, Rachel. 1994. *Silent Spring*. Boston: Houghton Mifflin. First published 1962.

Carson, Rachel. 1998. *The Edge of the Sea*. Boston: Houghton Mifflin. First published 1955.

Carson, Rachel. 2018. *Silent Spring and Other Writings on the Environment*. New York: Library of America.

Finch, Robert, and John Elder, eds. 1990. *The Norton Book of Nature Writing*. New York: W. W. Norton.

Gallagher, Mark. 2021. Live Deliberately, Stay Woke: Thoreau's Influence on William Melvin Kelley. In *Thoreau in an Age of Crisis: Uses and Abuses of an American Icon*, ed. Kristen Case, Rochelle L. Johnson, and Henrik Otterberg, 180–198. Paderborn: Brill Fink.

Giblett, Rod. 2008. *The Body of Nature and Culture*. Basingstoke: Palgrave Macmillan.

Giblett, Rod. 2009. *Landscapes of Culture and Nature*. Basingstoke: Palgrave Macmillan.

Giblett, Rod. 2011. *People and Places of Nature and Culture*. Bristol: Intellect Books.

Gore, Al. 1994. Introduction. In *Silent Spring*, ed. Rachel Carson, xv–xxvi. Boston: Houghton Mifflin.

Guattari, Felix. 1989. The Three Ecologies. *New Formations* 8:131–147.

Guattari, Felix. 2000. *The Three Ecologies*. Translated by I. Pindar and P. Sutton. London: Athlone.

McKibben, Bill, ed. 2008. *American Earth: Environmental Writing Since Thoreau*. New York: Library of America.

Steingraber, Sandra. 2018. Introduction. In *Silent Spring and Other Writings on the Environment*, ed. Rachel Carson, xix–xliii. New York: Library of America.

PART III

'The New Nature Writing' and Environmental Writing

CHAPTER 12

The Politics of People and Places in the City and the Country

12.1 Old and New Nature Writing

Richard Mabey stands like a colossus and acts as a bridge between the old nature writing as an editor, student, and proponent, and the new nature writing as a creative practitioner. Jos Smith begins his study of new nature writing by pointing in gushing tones to Mabey's 'monumental' *Flora Britannica* of 1996 as:

> a book that embodies some of the best qualities that have become characteristic of the New Nature Writing as a form, so much so that it has inspired an array of like-minded volumes on different topics. It is knowledgeable, innovative and inclusive; it relates to a long tradition of similar books while at the same time expanding what that tradition might be capable of; it looks at the plants, flowers and trees that are its subject with a sharp eye for local distinctiveness; and it meets the changed and changing conditions of the modern world around it with unflinching resourcefulness. Like many of his other books, it looks at the relationship between humans and wildlife as it has played out in the most diverse and distinctive ways. (Smith 2017)

Many of these characteristics apply to the old nature writing of Thoreau and White. This statement does not differentiate 'the New Nature Writing' from the old. It is hardly a definition of 'the new nature writing' and Smith is reluctant to provide one later too. Nearly a decade earlier, Jason

R. Giblett, *The Politics of Nature and Environmental Writing*, Transdisciplinary Environmental Humanities 1,
https://doi.org/10.1007/978-3-032-04427-3_12

Cowley attempted to differentiate and define them (which we will see shortly).

Smith then works backwards in time to create a genealogy for the new nature writing when he:

> makes a case for dating the emergence of the New Nature Writing to the early 1970s when, following the founding of Friends of the Earth UK in 1971, a new popular environmental movement began to spread via a fresh counterculture of activism and campaigning hitherto unheard of among the traditional conservation bodies. More precisely, I date the origins of this form to 1973 when, in the same year, both Raymond Williams's *The Country and the City* and Richard Mabey's *The Unofficial Countryside* were published, two books that would fundamentally change the way people thought and wrote about landscape in Britain. Williams drew attention to the political and economic realities embedded within, and structuring, a tradition of countryside writing, challenging a particular form of wealthy, metropolitan nostalgia that idealized rural life in ways that disavowed the reality of struggle for ordinary people. Mabey, in his book of the same year, used the conventions of countryside writing to, themselves, challenge the tidy border between country and city and turned a self-reflexive eye on the ambiguous terrain between the two that has come to be called the 'edgelands' since. (Smith 2017)

Smith does not associate the development of 'the new nature writing' from the 1970s with the rise of green politics and green political parties. The second wave of feminism from the 1970s too and the rise of ecofeminism with its critique of the gendering of nature do not rate a mention either. The old nature writing of John Muir was masculinist as it focused on the exterior landscapes of the earth (as we saw in a previous chapter). The old nature writing of Rachel Carson critiqued masculinism and militarism and created feminist political ecology (as we also saw in a previous chapter). 'The new nature writing' of Kathleen Jamie and Lydia Peelle is feminist and political as it focuses on the interior ecology of women's bodies. Both pieces were published in a special issue of *Granta: The Magazine of New Writing* devoted to 'the new nature writing' (Jamie 2008; Peelle 2008). Peelle's short story "Phantom Pain" is also about the interiors of animal bodies and an ageing, ailing male body. Jamie also writes about 'exterior' ecologies for *Granta* in "Airds Moss" and "Upriver," and she is highly critical of masculinist discourses on the outdoors in "Lone Enraptured Males."[1]

The Country and the City (Williams 1973) is not nature writing in the sense that White's *Selborne* or Thoreau's *Walden* is comprising observations of, and reflections on, nature, including as Williams puts it, birds, other animals, shapes of the land. By drawing attention to the aspects Smith highlights such as "political and economic realities," and critiquing "country writing" and the old nature writing, Williams' book can be credited with Mabey's with inspiring 'the new nature writing.' I credit *The Country and the City* with being the foundational text of ecocriticism and ecocultural studies (Giblett 2012, 2020a, chapter 10). The present volume is both political ecocriticism and an ecocultural study of nature and environmental writing that follows in its footsteps and those of Williams' later work.

Williams continued to contribute to these shifts and topics with his later work, such as his essay "Between Country and City" first published in a collection of essays edited by Mabey called *Second Nature* (and cited by Smith 2017). This essay includes Williams' own critical reflections on *The Country and the City* and his first discussion of livelihood, also developed in *Towards 2000* (Williams 1985, 1989). Williams acknowledged that his distinction between the country and the city in his book of this title suggested a hard and fast divide between them, whereas they are intertwined and mutually dependent in the livelihoods of everybody who dwells, lives, and works in either.

Smith does not discuss these later aspects of Williams' work, including his late essays on socialist ecology and the politics of place[2] (Williams 1989). Williams (1989, 237) points to "green socialism" and argues that it is "the most hopeful political movement of our time within which ecology and economics can become, as they should be, a single science and source of values, leading on to a new politics of livelihood [and] to a new social and natural order." This neglect of Williams' later political ecocultural work following on from the political ecocriticism of *The Country and the City* points to a recent and continual shrinking away from his radical politics and that of his contemporaries, such as E. P. Thompson, that can

[1] I am grateful to an anonymous reviewer for recommending the inclusion of these nuanced points and for drawing my attention to Jamie's other pieces in *Granta*.

[2] Smith lists Williams' interview about the politics of place in his bibliography, but does not cite it, nor discuss it, in the body of the book.

be found in both Smith's study of the new nature writing and the new nature writing itself in two pieces by Mabey and Macfarlane, and their disconnection from green politics (as I will argue in this chapter).

Smith then works forward chronologically to 2001 and credits Richard Kerridge with coining the term 'new nature writing' in a contribution to an American collection of essays pointing beyond the old nature writing and to pushing the boundaries of ecocriticism. Kerridge (2001, 138) argued that "environmentalism calls for a new nature writing, clearly differentiated from the conservative tradition [of the 'heavily tainted' interwar and post-war nostalgic English rural idyllism] and aware of its appeal and dangers." Kerridge did not say much more about 'new nature writing,' except to point to Thomas Hardy as the exemplar as he "offers the endless generation of meaning as the vivid life of a place, produced by its human, animal, and plant life" (Kerridge 2001, 138). Kerridge could have pointed to other instances for precisely the same exemplary qualities in the old nature writing, such as Thoreau. Kerridge is pointing to Hardy's novels which exclude him from the standard definition of nature writing, and so he is not included in the anthologies of nature writing (Finch and Elder 1990; Mabey 1995). By privileging Hardy, Kerridge seems complicit with nostalgic English rural idyllism.

Kerridge's appreciation for Hardy's writing in his novels is largely following in Williams' footsteps and couched in similar terms. Williams devotes an entire chapter of *The Country and the City* to an appreciation and discussion of Hardy's writing about nature in his novels, or what Williams calls "country writing" beginning with White (Williams 1973, 118). Williams regards Hardy as "the incomparable chronicler," and not as a natural or cultural historian (Williams 1973, 197 and chapter 18). The distinction between chronicler and historian is also crucial for Walter Benjamin in his discussion of storytelling and for Thoreau in *Walden* (Giblett 2025, chapter 7). Kerridge did not point, or could not point, to any contemporary examples of 'new nature writing' as it was either still to be born or in the process of being born.

Seven years later Jason Cowley took up the term 'the new nature writing' in 2008 in his introduction to the special issue of *Granta: The Magazine of New Writing* devoted to the topic and theme. Cowley attempts to differentiate the old nature writing from the new nature writing. The new nature writing for him consists of "voice-driven narratives told in the first person" with "the writer present in the story" that "experiments in form: the field report, the essay, the memoir, the travelogue" with “studies in the local or the parochial” involving “new ways of seeing” (Cowley 2008, 10). These are dubious and spurious grounds as the same could be said on each and every score about the old nature writing of Thoreau’s *Walden*, as well as recent environmental writing, such as Rebecca Solnit’s from the 1990s. Cowley does not mention the fact that all nature writing, both old and new, as well as environmental writing, also engages to greater or lesser extent with contemporary culture and nature, and with natural and cultural history, whether it be of a species, as in Mabey’s “The Tree of the Cross,” or of a place, as in Macfarlane’s “Ghost Species” (both published in the special issue of *Granta* devoted to ‘the new nature writing’), or many species and places as in White’s *Selborne* and Thoreau’s *Walden,* as well as Rebecca Solnit’s environmental writing. All nature writing, both old and new, as well as environmental writing, also engages to greater or lesser extent with local history, oral or written, or both as in these five cases.

Mabey’s focus on “plants, flowers and trees” in *Flora Britannica* continues with “The Tree of the Cross: In Search of the Fortingall Yew.” He situates his search for this individual tree within both natural and cultural history, including local history (oral and written) “historical ecology,” and “spiritual genealogy” (Mabey 2008, 95). Mabey is primarily concerned with the distant and recent past, not the present and the immediate future, nor with politics. He does not discuss the fact that ‘climate change’ is probably responsible for the severity and destructive impact of the hurricanes that blew down the Fortingall Yew in 1987 and again in 1900 (Mabey 2008, 95). Mabey cannot see the wood of climate change for the yew trees. Similarly Robert Macfarlane’s “Ghost Species” glosses over the history and politics of the natural and cultural environments of the Fens (discussed shortly). ‘The new nature writing’ in one way or another can’t see the wood for the trees. By focusing on the small scale, it overlooks the bigger picture of the larger scale and longer view of history and politics into the past and future.

The new nature writing points to British parochialism in the insular sense, both geographical and historical, though Cowley points to the recent American book and film *Into the Wild* with Christopher McCandless as a kind of latter day Thoreau and Muir who went off into the wilds of Alaska and ended up poisoning himself by not being able to identify a poisonous plant before consuming it (something that Thoreau would never have done). McCandless is a kind of exemplar for Cowley of the male Romantic hero that he associates with the old male nature writer, though McCandless went into the wild in the 1980s and published little and that largely posthumously. Cowley's brief mention of the old male nature writer is an acknowledgement of the gendering of the old nature writing, though Carson is the exception for her feminist political critique of masculinism and militarism. He does not acknowledge that all bar two (Jamie and Peelle) of the writers in the *Granta* special issue of new nature writing are male. The majority of the old nature writers are male, while some new nature writers are female. Neither Cowley nor Smith discusses the ecofeminist critique of the gendering of nature. McCandless is a kind of unfortunate throwback to a misguided male Romantic hero of a sort Thoreau never was, except for his excursion into the wilds of Maine and his two year sojourn by Walden Pond (Giblett 2025, chapters 1 and 9). He was as an astute student of flora and largely a creature of the parochial in the broader sense of place and the bioregion. Also unlike McCandless, Thoreau was sensible enough to have an indigenous guide with vast local knowledge when he went into the wilds of Maine.

12.2 Macfarlane's and Pullman's Forgetting of Fen Politics

Most new nature writers don't head off on excursions into the wild, but stay much closer to home in more homely and less remote, but still unfamiliar places. Robert Macfarlane's "Ghost Species" is a case in point as he begins by relating that he went on an excursion into the exotic place of the Fens not that far from home in search of a ghost and met some local farmers who have been living there for a long time. They and some plants and animals endemic to the place are becoming "ghost species." The emphasis on people means, though, that the distinctive features and history and geography of the place as a wetland and the politics of the drainage and enclosure of the Fens fade into the background in the past and in the landscape, and become merely the backdrop to the story about people in the present. Plants and animals are not agents and actors in this story. Macfarlane (2008, 111) writes:

> The Fens were once mostly water. In the ninth century, a Viking fleet could still sail as far inland as Ely. Until well into the 1600s, much of the Fens was a network of brackish swamps and reed beds, interspersed with islands and causeways of raised but marshy ground. The human inhabitants of this world were amphibious: travelling in punts, living in houses raised on stilts

> and surviving by fishing, cutting willow, reeds and peat, geese-keeping and wildfowling. In the 1620s, however, the Dutch hydro-engineer Cornelius Vermuyden, already renowned for his success in dyking the Thames at Dagenham, was employed by the Duke of Bedford to turn the Fens 'into a sea of waving corn'. So began the draining of the Fens with davit, dyke and windmill: a process that would take more than 150 years, but which revealed hundreds of thousands of acres of the most voluptuously rich soil in England. A soil so fertile, Fen farmers say, that if you stoop and scoop up a handful you'll grow three more fingers before you've cast it down. The Fens' terraqueous past is still visible in contemporary maps.

Macfarlane glosses over the politics of the draining of the "swampy" and "marshy" Fens (signalled in the weasel words "however" and "so began") and its wetland past presented in the Greco-Latinate abstraction "terraqueous." He also privileges the aristocrats' view of history by naming one such ducal actor in the draining of the Fens and by mentioning no resisters and dissenters, such as John Clare, the patron saint of the Fens who critiqued drainage and enclosure (Giblett 2021, chapter 4; 2024, chapter 6). The aristocratic and regal drainers were also enclosers of the commons, "a plain enough case of class robbery" as E. P. Thompson (a predecessor at Macfarlane's current place of employment at Cambridge University) called it 40 years before "Ghost Species" was published. The drainers not only met with resistance as drain destroyers were early Luddites against the enclosure of the common-wealth (Ash 2016, 2017). Macfarlane is on the side of the class robbers.

Macfarlane is of course not writing a historical geography of the Fens and its drainage. The history of the draining of the Fens has been told extensively elsewhere as documented by H. C. Darby (1956, 1974, 1983), another Cambridge predecessor and a historical geographer. Macfarlane, of course, was not writing a book about the Fens, which might have given him the opportunity to develop this discussion. Eric Ash (2016, 2017) has recently published extensive historical discussions of the draining of the Fens and I have published a critical study of the Fens' wetland cultures that discusses John Clare and other writers about the Fens, as well as painters (Giblett 2021, chapter 4; 2024, chapter 6). Recent books of 'the new nature writing,' such as Caroline Crampton's *The Way to the Sea: The Forgotten Histories of the Thames Estuary* and Rachel Lichtenstein's *Estuary: Out from London to the Sea*, develop and intertwine extensively the natural and cultural history, politics and contemporary culture and nature of the Estuary with their own story, including memoir and journeys (discussed later).

Macfarlane seems to be following to some extent in the footsteps of Philip Pullman in *Northern Lights*, the first part of Pullman's trilogy *His Dark Materials* published in 1995, that describes the Fens graphically as:

> that wide and never fully mapped wilderness of huge skies and endless marshland in eastern Anglia. The furthest fringe of it mingled indistinguishably with the creeks and tidal inlets of the shallow sea, and the other side of the sea mingled indistinguishably with Holland; and parts of the Fens had been drained and dyked by Hollanders. [...] But parts of it had never been drained or planted or settled at all, and in the wildest central region [...] eels slithered and waterbirds flocked, [...] eerie marsh-fires flickered and waylurkers tempted careless travellers to their doom in the swamps and bogs [...]. (Pullman 2011, 99)

No reasons are given for dyking and draining, and no consequences are drawn. They seem to be merely facts of history to note and pass over without comment and without attention to their class politics that slip into the background and forgotten.

By contrast, "Ghost Species" gives reasons for the draining of the Fens (to transform the wetland into dryland for agriculture) and draws consequences by indicating the benefits for agriculture of producing the most fertile agricultural land in England. On the other hand, "Ghost Species" does not draw the consequences of drainage for the harms it inflicted on the local paludiculture of the cultivation of wetland plants (peat, reeds, and willows) in the voluptuously fertile soil of the wetland already revealed to, and not hidden from, Fenlanders, for the animals they supported (fowl and fish) and for the people who farmed and ate fish and fowl and who were dispossessed.[3] "Ghost Species" reproduces the mainstream agricultural narrative and paradigm, largely ignores indigenous paludiculture, and neglects completely the radical, oppositional view that drainage and enclosure largely destroyed the indigenous Fenland economy and its livelihoods. Macfarlane does not bemoan nor mourn the loss of the wetland and its paludiculture, unlike John Clare whose writings, including on enclosure and the Fens, were studied by another predecessor at Cambridge University in John Barrell (1972). Wetland is not a word in Macfarlane's

[3] For further discussion of paludiculture, see Giblett (2024).

vocabulary in "Ghost Species" opting instead to use uncritically the loaded and denigratory terms "swampy" and "marshy."

"Landscape" is Macfarlane's general descriptor for the Fens and is used nine times in "Ghost Species" uncritically and without caution that this is a problematic term as documented by John Barrell (1972) and discussed by Williams (1973), another predecessor at Cambridge University. Smith's invocation of Williams as a founder of the new nature writing that "drew attention to political and economic realities" falls down at this point when considering "Ghost Species" as it does not do this. Barrell and Clare do not rate a mention in Smith's genealogy, nor does Clare and his writings. Macfarlane does not use the alternative term "wetlandscape" coined in 1996 and in circulation since then as a critical and decolonizing alternative to "landscape" in relation to wetlands such as the Fens (Giblett 1996). "Paludiculture" was coined in 1998 in relation to peat farming in northern Europe and has been applied recently to the Fens (Milner and Stuart 2022).

'The new nature writing' of "The Tree of the Cross" and "Ghost Species" forgets politics (and Smith's study follows suit), unlike the critical historical and radical political writing of Barrell, Thompson, and Williams (all Cambridge alumni luminaries), unlike American nature and environmental writing exemplified in Thoreau, Carson, and Solnit, and unlike the more recent 'new nature writing' of Caroline Crampton and Rachel Lichtenstein (discussed shortly). The 'new nature writing' of "Ghost Species" seems like a missed opportunity to consider the Fens, paludiculture, lost wetlands, and ghost swamps in the past and present landscape. By going on to focus on farmers, "Ghost Species" is farm writing, rather than nature writing, despite a late, desultory attempt to invoke a stray egret venturing out of its customary range due to climate change as a symbol of hope.

12.3 Recent 'New Nature Writing' of the Thames Estuary and Other Wetlands

'The new nature writing' continues to develop and has recently overcome some of its unfortunate birth defects and produced healthier progeny. Caroline Crampton (2019) in *The Way to the Sea: The Forgotten Histories of the Thames Estuary* and Rachel Lichtenstein (2017) in *Estuary: Out from London to the Sea* trace their journeys in whole or part down the

Thames River to, and into, its Estuary. The blurb for Lichtenstein's *Estuary* states:

> The Thames Estuary is one of the world's great deltas, providing passage in and out of London for millennia. It is silted up with the memories and artefacts of past voyages. It is the habitat for an astonishing range of wildlife. And for the people who live and work on the estuary, it is a way of life unlike any other—one most would not trade for anything, despites its dangers. Rachel Lichtenstein has travelled the length and breadth of the estuary many times and in many vessels, from hardy tug boats to stately pleasure cruisers to an inflatable dinghy. And during these crossing she has gathered an extraordinary chorus of voices: mudlarkers and fishermen, radio pirates and champion racers, the men who risk their lives out on the water and the women who wait on the shore.

Along similar lines, the blurb for Crampton's *The Way to the Sea* states:

> Raised on its banks and an avid sailor, Caroline Crampton sets out to rediscover the enigmatic pull of the Thames by following its course from the river's source in a small village in Gloucestershire, through the short central stretch beloved of Londoners and tourists alike, to the point where it merges with the North Sea. As she navigates the river's ever-shifting tidal waters, she seeks out the stories behind its unique landmarks, from the vast Victorian pumping stations that carried away the capital's waste and the shiny barrier that holds the sea at bay, to the Napoleonic-era forts that stand on marshy ground as eerie relics of past invasions. In spellbinding prose, she reveals the histories of its empty warehouses and arsenals; its riverbanks layered with Anglo-Saxon treasures; and its shipwrecks, still inhabited by the ghosts of the drowned. *The Way to the Sea* is [...] a fascinating portrait of an iconic stretch of water.

Rather than a landscape, or even a wetlandscape, *The Way to the Sea* is a portrait of the Thames and its Estuary.

Both *Estuary* and *The Way to the Sea* tick all the boxes to qualify as 'new nature writing' in Cowley's list of qualities for the genre, including mixing other genres. They are "voice-driven narratives told in the first person" with "the writer present in the story" that "experiments in form [combining] the field report, the essay, the memoir, the travelogue" with "studies in the local or the parochial" involving "new ways of seeing" (Cowley 2008, 10). Both *Estuary* and *The Way to the Sea* engage with local history, both oral and written. Lichtenstein has several pages of acknowledgements of people, films, books, and websites. Crampton also has several pages of acknowledgements of people and a select bibliography that also includes websites. Some of the people both of them mention include local environmental activists that make good on Smith's claim that they contributed to the development of 'the new nature writing.' Crampton and Lichtenstein also engage with the cultural and natural history of the Estuary and with the contemporary culture and nature of the Estuary, including of specific places and wetland sites located within the broader bioregions of the Thames Estuary and its catchments, and the politics of environmental conservation.

These specific places and wetland sites include: the present-day conservation of the Rainham Marshes (Crampton 2019, 162 and Lichtenstein 2017, 24); the current development threats to the Dickensian "marsh country" of Hoo Peninsula and the campaign to conserve it (Crampton 2019, 229, 233, 237–238, 241–242 and Lichtenstein 2017, 44–45, 221–228); and the rehabilitation of the Shell Haven oil refinery as the Stanford Wharf Nature Reserve (Lichtenstein 2017, 45) and Crampton (2019, 213). The Thames Estuary must be the world capital for the oxymoronic rehabilitation of rubbish tips and other toxic sites as "nature reserves," or as Crampton (2019, 26) puts it, "landfill heaps transformed into wildlife reserves." It is an irony of history that the first national parks were industrially useless or worthless lands (as we saw in Chap. 10), whereas nowadays industrially polluted wet wastelands in the Thames Estuary with which "there is nothing else to do" are being rehabilitated as "nature reserves," such as, for instance, Two Tree Island polluted with PCBs and DDT (Lichtenstein 2017, 109–110, 239–240). Both *Estuary* and *The Way to the Sea* are idylls of being on the water and spending time there messing about in boats when the land is not idyllic. Wetlands are caught in the middle.

By considering these clashes between the human and more than human worlds, Crampton's and Lichtenstein's 'new nature writing' is akin to North American nature and environmental writing that considers the collisions between, and impacts of, the human world of industrial capitalist technologies on the natural world of plants, animals, places, and people. These considerations began with steam railways in Thoreau's case, continued with artificial chemical ecocides in Carson's case and with nuclear bomb testing in Rebecca Solnit's case (as we will see in the following chapter), and culminated for the time being with pollution of land and water in Crampton's and Lichtenstein's cases and the incineration of migratory songbirds in the case of Harry Thurston and Thaddeus Holownia (2022). The old and new English nature writing of White's *Selborne*, Mabey's "The Tree of the Cross," and Macfarlane's "Ghost Species" concentrated on the natural world as a domain apart from the modern industrial world and did not consider the collisions and clashes between, and impacts of, the latter on the former.

The engagement of both *Estuary* and *The Way to the Sea* with the past, present, and future, especially with the presence of the past, as well as with the threats and opportunities of the future, suggests the fluidity of non-linear time. Lichtenstein concludes *Estuary* by writing:

> In the end, I feel I have created a kind of collective memory map, attempting to highlight what lies under the watermark, what has been obliterated, what is being obliterated and what is still under threat in this indefinable and beautiful place where past flows into present into past in its eternal rhythm. (Lichtenstein 2017, 320)

Crampton in similar terms writes in her "Epilogue" to *The Way to the Sea*:

> I have been many different versions of myself on this river, always returning to its shores to shed the shadows of a past self. But nothing we give to the water is ever really gone; it all ends up here, where the river deposits its secrets before its final transformation. (Crampton 2019, 296)

Nature writing old and new celebrates the rhythms of nature, the body, and autobiography, not just chronological time, and keeps memories alive of places lost and found, forgotten and remembered. The past is present.[4]

Black Swan Lake: Life of a Wetland (Giblett 2013a) tries to do so too. By tracing the life of the plants and animals of Forrestdale Lake through the six seasons of the local indigenous people, the first part of *Black Swan Lake* presents a wetlands calendar over a yearly cycle of the rising, falling, and drying waters of this internationally important wetland in south-western Australia. Forrestdale Lake is located on the outskirts of the city of Perth. It is not in the country; it is between the city and the country with suburban areas, battery hen farms, and horse agistments. It is not a wild place, but a feral place with introduced animals, such as foxes, and

[4] For further discussion of time, history, and memory along these lines drawing on the work of Thoreau and Walter Benjamin, see Giblett (2025, chapters 4, 5, and 7).

introduced plants, such as veldt grass. It is also a navigational point for light planes to land at Jandakot airport. They are an audible and visual reminder that this is not a wild place separate from industrial technology. It is also a natural place with native animals, including bandicoots and waterbirds, and native plants, such as banksias and orchids. It is a nature reserve and a wetland of international importance listed under the Ramsar Convention. Forrestdale Lake is located in the bioregion of the Swan Coastal Plain and is the confluence of the Swan-Canning River and Peel-Harvey catchments and the surface expression of their aquifers.

The second part of *Black Swan Lake* considers issues and explores themes from the first part, including a natural and cultural history of the seasons and the black swan. *Black Swan Lake* is a book of nature writing and environmental and cultural history and philosophy arising from living in a particular place with other beings. The book is a guide to living simply and symbiotically with the earth in troubled times and places by making and maintaining a strong attachment and vital connection to a local place and its flora and fauna. Local places and their living processes sustain human and other life on this living earth. I built my own house by Forrestdale Lake in 1986 and lived in it 28 years. For several years I kept a nature journal from which the first part of *Black Swan Lake* is drawn.

Like Thoreau, one of the founders of the American conservation movement, I built my own house by a body of water and wrote a nature journal. Unlike Thoreau, however, who lived by Walden Pond for a couple of years and kept a nature writing journal for more than 20 years which was published posthumously in its entirety, I lived by Forrestdale Lake for 28 years and I wrote a nature journal for a couple of years which was published in *Black Swan Lake*. I was also a local environmental activist for 25 years, including as the foundation Secretary of the Friends of Forrestdale and the writer of a book of local, oral history based on interviews conducted by Cath Drake (Giblett 2006). I combined local environmental activist, local historian, and nature writer. My activism and the writing that flowed from it engaged with politics and were politically green. Like Thoreau's sojourn and journal writing by Walden Pond which gave rise to *Walden*, the old testament of the conservationist's bible, my living and journal writing by Forrestdale Lake resulted in *Black Swan Like,* a book of both nature writing and environmental philosophy.

Similarly, like Leopold, another founder of the American conservation movement, whose renovation of a chicken shack in a sand county in Wisconsin and nature writing journal gave rise to the new testament of the conservationist's bible whose first and second parts are devoted to nature writing and the third part to essays in environmental philosophy, *Black Swan Lake* is divided along similar lines to *A Sand County Almanac* with the first part devoted to nature writing and the second part to environmental philosophy, natural and cultural history, and contemporary culture and nature (Leopold 1949). Like Leopold (1991), whose articles are collected in *The River of the Mother of God*, I have published articles in a wide range of journals and magazines and collected them in *Traces of an Active and Contemplative Life 1983–2013* and *Odds and Ends of a Writer's Life 2014–2019*, both published by Common Ground (Giblett 2013b, 2020b).

Lawrence Buell of Harvard University and author of *The Environmental Imagination: Thoreau, Nature Writing and the Formation of American Culture*, *Writing for an Endangered World*, and other books of ecocriticism wrote a cover endorsement for *Black Swan Lake* in which he said:

> *Black Swan Lake* is a unique synthesis of bioregional knowledge, historical and philosophical reflection, and autobiographical narrative with a distinctly personal stamp. I have learned a great deal from it and I am sure that other readers concerned with environmental ethics and the arts of environmental imagination, both within the academy and beyond, will feel the same.

Black Swan Lake qualifies as new nature writing in Cowley's list of qualities with the addition of natural and cultural history and engagement with contemporary culture and nature, environmental activism, conservation and politics. It is also a creative environmental art neglected in the usual roll call of the creative arts and a form of 'practice-led research' as *Black Swan Lake* indicates with chapters in the second part devoted to a culture history of the black swan, the colour black, the seasons, the history of Perth, and so on.

Buell concurs that nature writing is, or should be, more than about a sense of place and more about the processes of place and their workings within the wider context of the bioregion. The subtitle of Smith's study of the new nature writing is *Rethinking the Literature of Place* which suggests that he might hazard his own definition of place. He relies instead on Edward Casey who he says "has aptly described place as 'an ongoing cultural process with an experimental edge'." Nature writing is about place in these terms, and also about place as ongoing natural processes with an environmental and political edge. Nature writing can and should also be bioregional writing by situating a place or places within the broader context of a bioregion. A bioregion is a watershed and airshed in which people, plants, and animals live, which they create and on which they are dependent, especially people for their livelihoods in Williams' sense. The old nature writing of Thoreau's *Walden* and his other writing about Walden Pond and Woods is exemplary in all these regards. The focus on place in the new nature writing of Mabey's "The Tree of the Cross" and Macfarlane's "Ghost Species" neglects the bioregion in which the places they write about are located. The focus in the newer nature writing of Crampton's *The Way to the Sea* and Lichtenstein's *Estuary* on the Thames Estuary and its catchments locates specific places and wetland sites within their broader bioregions.

By developing a bioregional perspective in *Black Swan Lake*, I was only following in Thoreau's footsteps who is perhaps the writer to have most precisely and passionately expressed the sense of the bioregion as home-habitat. He wrote in his journal that "home is here" (Thoreau 1962, *V*, 254). He elaborated in his journal several years later:

> Here I am at home. In the bare and bleached crust of the earth I recognize my friend. [...T]he constant endeavour should be to get nearer and nearer *here*. [...] A man dwells in his native valley like a corolla in its calyx, like an acorn in its cup. *Here*, of course, is all that you love, all that you expect, all that you are. Here is your bride elect, as close to you as she can be got. Here is all the best and all the worst you can imagine. (Thoreau 1962, *XI*, 275; his emphases)

Home is here. It is not there. Life is here; it is not elsewhere. Home here is the bioregion. It is not the narrow confines of one's quarter acre, or fifth of an acre, suburban block, or one's townhouse and its pocket-handkerchief garden, or one's apartment and its balcony. One's bioregion is home; home is the bioregion. For Thoreau (1962 *VI*: 121) also "the earth is all alive."[5] For us humans, too, today, we need to (re-learn to) live with the earth as living being. Nature and environmental writing celebrates the earth as living being and gives guides for learning to live with the living earth.

[5] For further discussion of these aspects of Thoreau's thinking and writing, vital facets of psychopolitical ecology, see Giblett (2025).

References

Ash, Eric H. 2016. Reclaiming a New World: Fen Drainage, Improvement, and Projectors in Seventeenth-Century England. *Early Science and Medicine* 21:445–469.

Ash, Eric H. 2017. *The Draining of the Fens: Projectors, Popular Politics, and State Building in Early Modern England*. Baltimore, MD: Johns Hopkins University Press.

Barrell, John. 1972. *The Idea of Landscape and the Sense of Place 1730–1840: An Approach to the Poetry of John Clare*. Cambridge: Cambridge University Press.

Cowley, Jason. 2008. Editor's Letter: The New Nature Writing. *Granta: The Magazine of New Writing* 102: *The New Nature Writing*: 7–12.

Crampton, Caroline. 2019. *The Way to the Sea: The Forgotten Histories of the Thames Estuary*. London: Granta.

Darby, H. C. 1956. *The Draining of the Fens*. 2nd ed. Cambridge: Cambridge University Press.

Darby, H. C. 1974. *The Medieval Fenland*. 2nd ed. Newton Abbot: David and Charles.

Darby, H. C. 1983. *The Changing Fenland*. Cambridge: Cambridge University Press.

Finch, Robert, and John Elder, eds. 1990. *The Norton Book of Nature Writing*. New York: W. W. Norton.

Giblett, Rod. 1996. *Postmodern Wetlands: Culture, History, Ecology*. Edinburgh: Edinburgh University Press.

Giblett, Rod. 2006. *Forrestdale: People and Place*. Bassendean: Access Press.

Giblett, Rod. 2012. Nature is Ordinary Too: Raymond Williams as the Founder of Ecocultural Studies. *Cultural Studies* 26 (6): 922–933.

Giblett, Rod. 2013a. *Black Swan Lake: Life of a Wetland*. Bristol: Intellect Books.

Giblett, Rod. 2013b. *Traces of an Active and Contemplative Life 1983–2013*. Champaign, IL: Common Ground.

Giblett, Rod. 2020a. *New Lives of the Saints: Twelve Environmental Apostles*. Lanham, MD: Hamilton Books.

Giblett, Rod. 2020b. *Odds and Ends of a Writer's Life 2014–2019*. Champaign, IL: Common Ground.

Giblett, Rod. 2021. *Wetlands and Western Cultures: Denigration to Conservation*. Lanham, MD: Lexington Books.

Giblett, Rod. 2024. *Wetland Cultures: Ancient, Traditional, Contemporary*. New York: Palgrave Macmillan.

Giblett, Rod. 2025. *Waking to Nature with Thoreau and Benjamin: Or, Psychopolitical Ecology*. New York: Bloomsbury.

Jamie, Kathleen. 2008. Pathologies: A Startling Tour of our Bodies. *Granta: The Magazine of New Writing* 102: *The New Nature Writing*: 35–50.

Kerridge, Richard. 2001. Ecological Hardy. In *Beyond Nature Writing: Expanding the Boundaries of Ecocriticism*, ed. Karla Armbruster and Kathleen R. Wallace, 126–142. Charlottesville: University Press of Virginia.

Leopold, Aldo. 1949. *A Sand County Almanac and Sketches Here and There*. New York: Oxford University Press.

Leopold, Aldo. 1991. In *The River of the Mother of God and Other Essays*, ed. Susan Flader and J. Baird Callicott. Madison: University of Wisconsin Press.

Lichtenstein, Rachel. 2017. *Estuary: Out from London to the Sea*. London: Penguin Books.

Mabey, Richard. 1995. Introduction. In *The Oxford Book of Nature Writing*, ed. Richard Mabey, vii–x. Oxford: Oxford University Press.

Mabey, Richard. 2008. The Tree of the Cross: In Search of the Fortingall Yew. *Granta: The Magazine of New Writing* 102: *The New Nature Writing*: 86–96.

Macfarlane, Robert. 2008. Ghost Species. *Granta: The Magazine of New Writing* 102: *The New Nature Writing*: 109–128.

Milner, Jim, and Judith Stuart. 2022. Paludiculture—The Future of Farming on Peat Soils? https://naturalengland.blog.gov.uk/2022/09/30/paludiculture-the-future-of-farming-on-peat-soils/. Accessed 3 May 2023.

Peelle, Lydia. 2008. Phantom Pain. *Granta: The Magazine of New Writing* 102: *The New Nature Writing*: 131–151.

Pullman, Philip. 2011. *His Dark Materials*. London: Random House.

Smith, Jos. 2017. *The New Nature Writing: Rethinking the Literature of Place*. London: Bloomsbury.

Thoreau, Henry David. 1962. In *The Journal of Henry D. Thoreau, Volumes I—XIV*, ed. Bradford Torrey and Francis Allen. New York: Dover.

Thurston, Harry, and Thaddeus Holownia. 2022. *Icarus, Falling of Birds: A Poem by Harry Thurston, Photography by Thaddeus Holownia*. Jolicure: Anchorage Press.

Williams, Raymond. 1973. *The Country and the City*. London: Chatto & Windus.

Williams, Raymond. 1985. *Towards 2000*. Harmondsworth: Penguin.

Williams, Raymond. 1989. In *Resources of Hope: Culture, Democracy, Socialism*, ed. Robin Gable. London: Verso.

CHAPTER 13

Landscape Politics and the Exquisite: Rebecca Solnit

Rebecca Solnit (2003, 2, 5, 9; 2007, 123, 307) has described herself variously as a writer, critic, historian, environmental journalist, and environmental activist who has worked as an art critic and land rights activist. Her politics are progressive and she puts them into practice as an activist by protesting against nuclear testing, advocating for what she calls "global justice" (Solnit 2007, 174) in "the occupy movement" (Solnit 2014b, 219), and marching to reclaim the night and the streets as safe times and places, especially for women and people of colour. Her progressive politics occurs under the overarching umbrella of the "rainbow coalition" for equal rights and social justice for all, irrespective of sexuality, gender, place, "race," ethnicity, ability, colour, class, or creed. She contributes many vital facets to progressive political ecology.

Solnit also puts her politics into practice as the writer of politically progressive books, essays, and columns where the pen counters the might of the sword through the power of critique and storytelling, including recently retelling the classic children's story/fairy tale of Cinderella (Solnit 2019a). Her writing creatively crosses the great divide between fiction and "non-fiction," "the leftover term apotheosizing fiction" as Solnit (2014b, 4) calls it, as well as denigrating the essay, her preferred genre for affirming places (topoi) and topics (Solnit, 2014b, 4).

Solnit has been particularly adept at exposing and excoriating the double standards of racist and sexist patriarchy. She has been credited for

R. Giblett, *The Politics of Nature and Environmental Writing*, Transdisciplinary Environmental Humanities 1,
https://doi.org/10.1007/978-3-032-04427-3_13

inventing the concept of "mansplaining" (Solnit 2014a). She has been called "the voice of the resistance." Her essays have been published most recently in *The Guardian* daily online newspaper and weekly print magazine. As I was writing this chapter, the latest *Guardian* weekly magazine that reprinted a stirring and timely essay by Solnit (2020a) from an earlier daily edition of *The Guardian* arrived in my mailbox. It drew together and hit the spot on COVID-19, bushfires, climate change, Trump, politics, and hope. After the infamous and failed Capitol insurrection of January 6, 2021, the daily edition of *The Guardian* published her truth-telling essay calling it out as an attempted coup (Solnit 2021). Her essays are usually later collected into books (Solnit 2014a, 2017, 2018, 2019b, 2025).

Solnit has been canonized by a jury of her readers and peers as St Rebecca of San Francisco, the living "patron saint of progressive politics," embracing feminist politics, indigenous American politics, African American politics, landscape politics, and the politics of landscape photography. Her wide-ranging writing operates in the broad intersection of these and other avenues of inquiry and intervention. My cursory and introductory discussion to the vast range of her work only touches its surface and invites you, the reader, to delve into its depths.

13.1 Landscape Politics and Photography in Yosemite

Whatever topic or topos Solnit is tackling, whether it is walking, photography, landscapes, national parks, native peoples, nuclear bomb-testing, American cities, or the aftermath of disasters such as the aftermath of Hurricane Katrina in New Orleans, it always involves her undertaking extensive historical research in the archive and fieldwork on the ground. "The past is equipment for the future," as Solnit (2014b, 234) puts it succinctly. Tackling the politics of landscape photography is no exception as it means for Solnit researching its history from its beginnings in the nineteenth century. The American photographer Carleton Watkins for Solnit (1999, 237; 2004, 45) is "the founder of the major photographic tradition of landscape on this continent" and "the most widely acclaimed landscape photographer of his day." His most famous photographs of the sequoias and Yosemite were "virtually the first major landscape photographs in the United States," Solnit (2003, 45) argues, if "landscape" is defined as the aestheticized surface of a piece of the land seen as scenery in the modalities of the sublime, beautiful, or picturesque. Watkins' photographs were taken during the Civil War and were instrumental in having Yosemite "set aside" during the war as "the world's first national park," as Solnit (1999, 221) suggests. It was decidedly an *American* national park as "Yosemite had become a national landscape." It was "a great national symbol of nature" as "the United States rooted its identity in its landscapes," as Solnit (2004, 44–45) points out.

Yet Yosemite was not only set aside during one war but also the result of a previous one. To become a national park it had been a battlefield in the war against the local indigenous owners and inhabitants in 1851 from which the losing side was progressively removed, the history of this war erased, their place names violated, their "sense of place" obliterated, their work in creating the park by their labour in the first place appropriated, and their presence in the past and present unacknowledged (Solnit 1999, x, 29–30, 219–220, 257, 279, 304–308). Yosemite, as Rebecca Solnit (1999, 222) argues, "was a battleground before it was a vacation destination." Before it was a battleground, it was a garden that resembled an English landscape garden, because it *was* a garden created by the local indigenous people (Solnit 1999, 257, 307). The same sad and sorry story is told in Australia too (see Giblett 2011, 91–94).

In the period between being a battleground and a vacation destination and going from "an indigenous stronghold to a war zone to a tourist attraction," as (Solnit 1999, 232) traces the transformation, Yosemite had become a work of art. In fact, to become a vacation destination and tourist attraction, it first had to become a work of art. Yosemite, as Solnit (2007, 28–29) puts it, "would become the great national shrine of nature as a work of art." The US was "nature's nation," as Perry Miller called it, with Yosemite as its pinnacle as a work of art and as a shrine in its national natural religion. It also became internationally important, not only as the archetypal national park but also as a tourist attraction. Yosemite National Park as Solnit (2000, 149) argues is "one of the most famous and heavily visited natural places in the world," though it is hardly a purely "natural" place, as Solnit is it pains to point out elsewhere, but a cultured place. Before it was visited, it was photographed, and because it is visited, it is photographed. Solnit (1999, 221) also argues that Yosemite "for landscape photography is one of the most important places in the world." No place on earth for Solnit (1999, 222) "is more central to landscape photography and landscape preservation." The national park is a landscape preserve, rather than a conservation reserve. It was usually and initially set aside for aesthetic reasons and/or touristic purposes as Yosemite demonstrates.

Solnit (1999, 232) suggests that "the birth of national parks, of landscape photography, of the conservation movement" began with, and took place in, Yosemite. More precisely, the sanctuarist branch of the American conservation movement under John Muir's leadership did that by preserving special places in sanctuaries, rather than the sacrality side of American conservation before it became a movement under Thoreau's guidance that advocated conserving sacred nature everywhere (see Giblett 2011). For Solnit (2007, 92), "setting gardens apart is how the conservation movement began," whereas American conservation as such began with Thoreau setting dismal swamps as sacred places apart from the manicured lawns of urban gardens.

Muir for Solnit (1999, 259) was "probably the greatest voice for nature as a religion [...] who spoke as a virtual evangelist of wild scenery and who found in Yosemite his temple." The same could probably be said of Thoreau, except that he found his temple in a swamp and he did not have Muir's audience in his lifetime to be as great a voice for nature as a religion as Muir was. Unlike the safe Muir, "the dangerous Thoreau" as Solnit (2007, 189, 192; 2014b, 274) calls him was "an outspoken antiwar activist, tax resister and eco-anarchist" as well as "our nation's or empire's trickster-in-chief." Thoreau was also a nature writer, contributor to a journal of small circulation (his own), cultural botanist, land surveyor, graphite pencil inventor, own-home builder, self-appointed snow-storm inspector, sauntering walker, swamp lover, slavery abolitionist, underground railroad worker, track-repairer somewhere in the orbit of the earth, anti-state thought plotter, berry-picking party leader, weaver and purveyor of textual baskets, immanent philosopher, and pantheistic theologian. The rich variety of all these aspects of Thoreau the thinker, writer, and worker go hand-in-hand with each other.

Muir's "wilderness writings" for Solnit (2000, 126) made him into "a kind of John the Baptist. [...] Muir is the United States' evangelist of nature." Indeed, he has become canonized as St John the Baptist of the Sierras, "the patron saint of national parks," a voice crying in the wilderness following in the footsteps of the old testament prophet-poet of Thoreau and preparing the way for the messiah of Aldo Leopold and his saving gospel of a land ethic and a conservation aesthetic. Yosemite for Muir was a cathedral, a rocky outcrop his pulpit, a pool his baptismal font or bath, for delivering his message and converting citizens to conservation, just as he himself had been converted by his encounter with an orchid in a Canadian marsh and an alligator in a Florida swamp. His sanctuarism of Yosemite was tied up with the contemporary secular aesthetics of the sublime.

Yosemite, Solnit (1999, 263) argues, "has been preserved as though it were a painting." It has not been preserved as if it were a battlefield, though a battlefield should be regarded as a landscape. Landscape, for Solnit (2003, 80), "should have been revised a long time ago to include battlefields." Battlefields for Solnit (2003, 10) are "one kind of landscape." Landscapes like Yosemite are also battlefields. Photography of battlefields began in the same period as photography of Yosemite. If landscapes should include battlefields, then virtually the first major landscape photographs in the US were both of the battlefields of the Civil War taken by Matthew Brady and Timothy O'Sullivan and of the battlefield of the aftermath of "the Indian War" in Yosemite taken during the Civil War by Watkins.

The landscape of battlefields shifts historically from France as feral swamp on the Western Front in World War I to Germany as urban ruin in World War II that marks a profound, elemental shift in the exercise of armament, in the deployment of firepower and its destructive force, and the experience of war for civilian and soldier citizens.[1] They also mark a shift in the type of landscape produced: from feral swamp to cities "ruined by war" (Solnit 2005a, 89); and from the human-made countryside remade into the human-made swamp to the human-made city remade into the human-made ruin. A ruin is defined by Solnit (2005a, 88) as "a human construction abandoned to nature." A bombed city is an artificial ruin, a human construction destroyed by culture. A feral swamp is a native construction ruined by humans. Both swamp and ruin stand as figures for the unconscious. Just as the swamp is a figure for the cultural unconscious (see Giblett 1996), so ruins for Solnit (2005a, 89) "become the unconscious of the city." Out of both new life can spring as Solnit (2005a, 90) argues that "with ruin a city comes to death, but a generative death like the corpse that feeds flowers," just as the corpses did in the battlefields of World War I. Ruins for Solnit (2007, 355) are "evidence not only that cites can be destroyed, but that they survive their own destruction, are resurrected again and again" as bombed German cities show.

[1] For further discussion of the landscapes of World Wars I and II, see Giblett (2009, chapters 4 and 5).

Landscape may not include battlefields, but landscape photography has included battlefields for a long time, as demonstrated in its beginnings from Matthew Brady and O'Sullivan and the American Civil War, and in its subsequent history through the photographs by the Australian Frank Hurley of the western front during World War I, through Richard Misrach's of the Cold War, to Joel Meyerowitz's photographs of the site of the former World Trade Center in the aftermath of "9/11" and the War on/of Terror.[2] Landscape photography lends itself to depicting the monumental, whether it be the massive rocks of national parks, or the massive devastation of battlefields, nuclear test sites and demolished skyscrapers, and their aftermath. Satellite photography does so too. Most of the debris from the destruction of the World Trade Center on "9/11" was transported to "Fresh Kills Land Fill" on Staten Island which has "exceeded the Great Wall of China as the largest manmade object on earth" (Solnit 2003, 53 and 149), a dubious distinction.

Rather than landscape needing to be revised to include battlefields as it had already included them, considerations of landscapes from an aesthetic and historical point of view need to be revised to include them. From the military point of view, landscapes of the battlefield have been an important part of war for a long time with the development of aerial photography; from the art history point of view, the battlefield has not been an important aspect of landscape for long, yet from the landscape photography point of view, photography of battlefields has long been included; from the critical cultural history point of view, battlefields, bomb-sites, national parks, wastelands, cities, and swamps are all landscapes.

Even when landscape photography did not include battlefields, such as in Ansel Adams' landscape photography of Yosemite National Park taken long after it had ceased to be a battlefield, the World War II and the Cold War were the backdrop and the rationale for his monumental and comforting photographs of this quintessential, nationalist American landscape. Yosemite National Park is "the very crucible and touchstone for American landscape," as Solnit (1999, 221) puts it. Just as the Civil War was the nationalist impetus for the declaration of Yosemite as a state park and for the national park movement, and national parks were the place Adams depicted in World War II as patriotic landscapes, so was the Cold War the driving nationalist force behind the wilderness movement. The American

[2] For reproductions of photographs by most of these photographers, see Giblett and Tolonen (2012, 103–151).

wilderness was the American domain where the American scene could be seen and contemplated in all its glory and where the American idea of wilderness and itself could be maintained.

Just as the US sought and found a comforting and unifying sense of national identity in Yosemite during the Civil War, so it found it there again during World War II and the Cold War. It was a bulwark against, and consolation for, the threat of thermonuclear war with the Soviet Union. The wilderness was also the place where both the might of America was displayed and the land wasted in the testing of nuclear bombs. America displayed its power both to conserve lands by setting aside pieces of and to destroy it by wasting pieces of it elsewhere. Rather than being conservation sites, both national parks and wilderness are landscapes of warfare because they are places where the nation sought solace and its national identity against the threat of war and where it could display its national vigour and power. Adams is the photographer of the nationalist sublime par excellence.

Adams is Adam in the Garden of Eden of landscape photography who acknowledged his predecessor in Watkins. The Sierra Club was the vehicle to do so. Adams was a long-time member of the Sierra Club founded by Muir and friends in 1892. Solnit (2000, 154) calls it "the first major force for environmental protection in the United States." The aesthetic was one means to do so. Not just a preservation organization, the Sierra Club for Solnit (2007, 235) "put the aesthetic to political use in a way no other environmental group had" before and after. Most environment groups have done so ever since, such as the Australian Conservation Foundation with its diaries and the Australian Wilderness Society with its calendars. Print was the medium for the aesthetic to put to political use with photographic calendars and books that, as Solnit (2007, 251) points out, "first made the Sierra Club a visible force nationwide." America with Adams' photographs, and its national parks, and wilderness, could have its conservation cake and eat its nature cake as well. In other words, it could conserve "vast tracts of wilderness" and set aside national parks while ruthlessly exploiting and polluting nature everywhere else in industrial agriculture and mining. It preserved, painted, and photographed nature in Yosemite National Park and bombed it just over the Sierra Nevada mountain range in the "Nevada Test Site" in about the same latitude (as the map in Solnit 1999, viii–ix shows).

Both places are landscapes of war with Yosemite National Park as the site of the war against its indigenous owners and inhabitants; both place-scapes are also time-scapes, as nuclear war was "supposed to be our future" while "the Indian wars" were supposed to be "of our past." Solnit (1999, x, 220) argues of indigenous that "there is no room for these people in the present, but they will become a decorative past for someone else's future." Colonization is just as much about time as it is about space, place, and land. Colonization colonizes past, present, and future in chronological time. Decolonization decolonizes what Solnit (2009, 165) calls "the linear time of progression" into "the cyclical time of celebration" in the carnival (Solnit 2009, 165–169), or in more arcane terms, chronological time (or *kronos*) into the *kairos* of the eternal and messianic now, such as the cycle of the seasons. Carnival for Solnit (2014b, 246) is "inherently against capital." It is against the market and for the marketplace, the place of the carnivalesque.

13.2 Landscape Politics and Photography in Nevada

The Nevada rehearsal space for thermonuclear war is the site of what Solnit (1999, 30, 86) calls "military and industrial assaults" against the natural and cultural environments of Nevada as "the US government had begun a war against the people and land here long ago." The US government had also begun a war against its own citizens, the land, the air, and the water in the greater US as all Americans are "downwinders" (Solnit 1999, 42) who experience nuclear fallout in some way, shape or form, including in their bones and food, as Solnit (1999, 97) points out. She concludes that "the instruments of the Cold War, the bombs themselves, were endangering the people they were supposed to protect" (Solnit 1999, 99).

The tying together of American nature and nation in Adams' work becomes unravelled in the work of the "explicitly activist" Atomic Photographers Guild who were murderous sons who killed their father in Adams, who is for Solnit (2003, 91) "the Great Oedipal Father of American Landscape Photography." The Atomic Photographers also re-established an alternative lineage (on their mother's side?) back to the beginnings of "American landscape photography" in 1867 when Civil War photographer Timothy O'Sullivan "went west" (Solnit 2004, 40) and produced photographs of what Solnit (2003, 66) calls "an array of weird formations and geological wonders." O'Sullivan's photographs hang like a ghost over the work of the Atomic Photographers who also followed the recent alternative lineage of the "New Topographics," not merely for the clarity and precision of his depiction of the deserts of the American West, but also for his invocation of the uncanny underside of the earth. The Atomic Photographers saw and photographed the results of the testing of nuclear bombs in the Marshall Islands and at the "Nevada Test Site."

Located 60 miles north of Las Vegas, the "Nevada Test Site" is described by Solnit (1999, 208, 378; 2003, 74; 2007, 1, 128) as "the most bombed place on earth" with 1000 nuclear bombs detonated there during "forty years of nuclear war on the land" between 1951 and 1991 making it "the world's biggest—and maybe dumbest—nuclear waste dump." Rather than its official name of "test site," Solnit (1999, 4, 5, 15) describes the Nevada "rehearsal site" as "the place where the end of the world had been rehearsed since 1951." Rather than making "nuclear war a terrible possibility," these rehearsals made the rehearsal site into "an ongoing regional catastrophe" as Solnit (2007, 128) points out later. They also made the rehearsal period into an omnipresent global threat. These rehearsals "shook the earth" and created "limited earthquakes" with "a rippling of ground motion," as Solnit (1999, 16; 2000, 7) points out elsewhere. The rehearsal site was a feral quaking zone made with human hands. The surface of the earth was "pockmarked with depressions [...] that looked [...] like the devastated skin of an ancient plague survivor," as Solnit (1999, 207) puts it later. The body of earth was bombarded with radiation and the symptoms were manifested on its surface.

Solnit (1999, 44) finds the power of the bomb "fascinatingly terrible" and associates it with the sublime. The bomb itself demonstrates, as Solnit (43) puts it, that "humans have managed to make stars," but "an exploding nuclear bomb is a kind of star come to earth," with catastrophic and uncanny consequences as Richard Misrach's photographs show. Misrach for Solnit (1999, 39) is "the pre-eminent landscape photographer in the country," partly because "the unnatural disaster is his most persistent theme" as she puts it (Solnit 2003, 96).[3] The destructive force of the bomb is "fascinatingly terrible," a shorthand definition of the sublime,

[3] For further extensive discussion of Misrach's work, see Solnit's essays entitled "Scapeland" (Solnit 2003, 63–89) and "Excavating the Clouds" (Solnit 2007, 143–164). See also her comparison of Misrach photographs with Edward Burtynsky's (Solnit 2007, 137–138). A scapeland is not only a reversed landscape, or an anti-landscape, but also a scapegoat, not only deserts as disposable bomb sites with nuclear wastes, but also wetlands as places where sins are driven out, heaped, and expiated.

and its aftermath fascinatingly horrible, a shorthand definition of the uncanny. The bomb encapsulates the two ruling passions of modernity: terror and horror—the terror of the explosion and the horror of mutilated bodies and lands. Misrach's photographs for Solnit are the fascinating synthesis of both the horrible and the sublime. Solnit (1999, 46, 47) sees Misrach pursuing in his photographs "this new, horrible, postmodern sublime."

The bomb could not remain a star in the heavens but came to earth in the hypermodern sublime of the nuclear bombing. Solnit (1999, 207) describes aerial views of the bombing in which "the earth shuddered [...] as though it were made of thick liquid. [...] Aerial photographs of the Ground Zero region show an arid surface pockmarked with depressions and crisscrossed by roads like long slashes, a surface that looks more like the devastated skin of an ancient plague survivor than the familiar surface of the earth." The earth becomes a quaking zone made with human hands and the surface of the earth is inscribed with the marks of a modern plague against the earth and its inhabitants.

Crisscrossing roads slashing the surface of the land morphed into the humungous freeway where monumental constructions were erected into visible shrines to national identity that replaced the national park. Roads for Solnit (1999, 365) are "the real architectural achievement of this [twentieth] century in the U.S., freeway overpasses and cloverleafs the greatest monuments to its beliefs." In a slightly different register, she goes on later to suggest that "roads are the greatest feats of American building, and the graceful clusters of pillars that support freeway ramps will come to be seen as its temple architecture. [...] Roads are not only the major accomplishment of the American spirit of manifest destiny [to conquer the American West and the world] but also its principal expression" (Solnit 2003, 185). Just as the autobahn is an icon of Nazi Germany, so the freeway is a monument to American civic and corporate exceptionalism in landscape design, engineering, and construction.

13.3 Nature Photography and the Exquisite

Aerial views of crisscrossed roads, looping cloverleafs, and the pockmarked surface of the earth are a far cry from the smooth surface of the earth in nature photography. Ansel Adam and his successor Eliot Porter for Solnit (2000, 153) are "the men who invented American nature photography" as a viable commercial enterprise and central cultural institution.[4] Solnit (2003, 200) takes aim at both of them and their successors in drawing a comparison between nature photography and pornography in an essay appropriately entitled "Uplift and Separate." She notes the similarities between nature photography and pornography as "both are souped up, sleek, passive [...] in which the chaos of thought and action have been replaced with a stale vision of delight." She goes on to enumerate seven features of the "nature photography genre" (Solnit 2003, 201–202), or nature pornography genre, or "eco-porn":

1. No human beings or their traces—that is to say, no history:
2. Nothing dead, sick, rutting, dying, or a in state of decay—that is to say, no natural history [...];
3. Water's main purpose is to mirror, with glasslike perfection, the landscape looming above it [...];
4. Repetition and pattern are good [...];
5. Colours should be bright [...];
6. All animals are lovable and attractive [...];
7. The photograph itself should be so clean as to never call attention to its own creation, but rather to *Creation*.

In resistance and opposition to "nature photography," what Solnit (2007, 138) calls "a truly ecological photography" would trace "the life of a commodity from extraction to disposal." Yet tracing the life of a community in its place and in its use of living beings before they become, or when they are not, or have ceased to be, a commodity might portray livelihood in what could be called "photography for environmental sustainability" (Giblett and Tolonen 2012, chapter 13).

Also in resistance and opposition to "nature photography," Solnit's (2003, 207–215) proposes the Exquisite as "a new aesthetic category" for "the aesthetic of unimproved conscious female bodies" (Solnit 2003, 207). The Exquisite for Solnit (2003, 211) "revels [...] in the transgressible boundaries of the female, the liminalities or permeabilities that are denied in the masculinized ideal body," especially, I would add, in the fascist body (see Giblett 2008), the cold, hard, impermeable body. The exquisite revels also in the transgressible boundaries of the female

[4] Solnit (2007, 225–253) devotes a long essay to discussing extensively Eliot Porter's photography.

commons of the wetland, in the liminalities between land and water, or in the permeabilities between solid and liquid that are denied in the masculinized sublime city and found in the feminized slimy swamp. In addition, the Exquisite for Solnit (2003, 212) "affirms the biological body against prior artistic norms" and "restores [...] the squishy, mobile, mutable stuff of bodies in process." The exquisite also affirms the body of the earth against the prior artistic norms of depicting (or not) wetlands, and restores the same messy stuff of the body of the earth in the processes of birth, life, death, and regeneration in slimy swamps. Moreover, the Exquisite for Solnit (2003, 212) "operates in this zone where the repulsive and the beautiful intersect, or rather in the zone where what is conventionally considered most beautiful—the female body [or water body]—and what is conventionally considered most repulsive—also the female body [or swamp or marsh]—intersect." The Exquisite also operates in the native quaking zone of wetlands. Thus the Exquisite for Solnit (2003, 214–215) is "an aesthetic in which the beautiful meets the repulsive, insofar as the damp messes and tangles of biology are repulsive." The Exquisite also operates in the quaking zone between the beautiful and the repulsive damp messes and tangles of marshes and swamps and their biological processes. Finally, the Exquisite for Solnit (2003, 215) operates in the psychological realm of the Abject. The Abject also operates in the anti- or counter-aesthetic realms of the uncanny, grotesque, and monstrous, so the Exquisite as an aesthetic category is also linked to anti- or counter-aesthetic ones via the Abject.

A new embodied environmental aesthetics would involve a sense of place for this sense is for Solnit (1999, 203) the sixth sense. This sixth sense of place is topaesthesia as well as love of place, or topophilia. Sense of place for Solnit (2013, 30) leads to, or is more precisely and affectionately, "love of place," not only "our love for places," but also "how [...] places love us back" and appreciating "what they give us." From the get-go, Solnit has been a writer in, for, and about place. Over the course of her career, she has developed and refined her understanding of what place is, how to define it, and how to live in it. Place for Solnit (2014b, 1) is "the intersection of many changing forces passing through [...] a fixed location." In a place "ecology, democracy, culture, storytelling, urban life, individual life histories and collective endeavour coexist geographically, spatially." Solnit (2014b, 3) celebrates "the pleasures and possibilities each place holds [...] with the beauties of light, space, and solidarity."

13.4 Cities Such as New Orleans and Paris

Landscapes and their politics for Solnit include places such as battlefields, national parks, nuclear "test sites" and cities. Beginning with her own home city of San Francisco (Solnit 2010) and then with various collaborators, she has produced a stunning trilogy of cultural atlases of three American cities, including New York (Solnit and Jelly-Schapiro 2016). In the second of these atlases devoted to New Orleans, Solnit and Snedeker (2013, v) describe the city as "resting on the pillowy softness of river-delivered muck [and] mud" and "at the center of the American unconscious." New Orleans for Solnit and Snedeker (2013, 1) is "a city of amorphous boundaries, where land is forever turning into water, water devours land, and a thousand degrees of marshy, muddy, oozing in-between exist." As a result, for them, "all that is solid dissolves into water," much of which "seems to exist in an amorphous state of muddiness and murkiness" (Solnit and Snedeker 2013, 2). All that is solid dissolving into water is the process of desublimation, while the counter to sublimation in which all that is solid melts into air.

New Orleans is "the birthplace of jazz and the capital of the Mississippi Delta," Solnit (2009, 271) says. The delta is not only site of an iconic American city but also the cloaca of the continent. Solnit and Snedeker (2013, 82) describe how:

> way down where all the effluvia of the continent drains out, all the toxins and manure and muck of a great river system flowing through agricultural and industrial lands, way down there underneath in the softness of the alluvial soil, is the cloaca or pudendum of the continent. [...] 'Pudendum' a word for the female genitalia, comes from a Latin word for shame.

'Cloaca' is also a word for a single opening for excretory and reproductive functions in some animals, also a sewer and gathering place of moral evil, in other words, a sump of iniquity. Rather than the swamp being shameful in its native and pre-contact state, it is the polluted river and swamp cloaca that is shameful in its present feral state.

Continuing along these lines, Solnit and Snedeker (2013, 2):

> think of New Orleans as a liver, an expanse of soggy land doing some of what a liver does, filtering poisons, keeping the body going, necessary to survival and infinitely fragile, hard to pull out of context, and nowadays deteriorating from more poisons than it can absorb, including the ongoing

toxins of the petroleum industry and the colossal overdose delivered by the 2010 BP blowout.

The liver "is delicate, spongy tissue" and it can "shred" when it is operated on and "rip" when it is stitched (Solnit and Snedeker 2013, 2). Similarly around New Orleans "the land is soft and marshy" and industry is "ripping and shredding the land into nonexistence." The distinction between the living and the dead was the hardest to maintain. "Just as New Orleans is a place of unclear boundaries between land and water," for Solnit and Snedeker (2013, 34), "so it is a place where the boundaries between life and death are thin." As a result for them, it is "a city of the living that is also a city of the dead."

This state of affairs was quite literally the case in the aftermath of Hurricane Katrina when the city itself was also dying and living. New Orleans became what Solnit (2009, 1, 2) calls "the septic, drowned city" with its "stagnant waters." "The most hellish image in New Orleans" for Solnit (2005c) was "the forgotten thousands crammed into the fetid depths of the Superdome," "the stinking, scorching zone" (Solnit 2014b, 233). The human-made swamp, the feral quaking zone both inside and outside the Superdome, had worse conditions of air, water, and sanitation than any swamp not made with human hands, any native quaking zone. New Orleans in the aftermath of Katrina was also a place of citizen activism and "mutual aid" as Solnit (2009, 86) relates in an extensive case study of one of "the extraordinary communities that arise in disaster" as well as "perhaps the worst case of elite panic in the history of the United States" (Solnit 2009, 235; see 231–304, 311–312). Mutual aid implies "cooperation rather than competition," compassion and reciprocity rather than charity and empathy as Solnit (2009, 86, 95, 118) puts it. Unimpaired mutual aid creates civil society for Solnit (2009, 146). "New Orleans was saved by love," Solnit (2014b, 238) concludes in a retrospective essay written five years after Katrina.

Katrina is also an instructive case study in how a disaster can become a catastrophe. It began with what Solnit (2009, 234, 239–240) describes as "the somewhat natural disaster of the storm [and] the strictly unnatural disaster of the failing levees, a man-made disaster of levee collapse and urban flood." It became a succession of disasters that evolved into what Solnit (2009, 240) calls "a socio-political catastrophe" of "the hot, filthy, ruinous city." Solnit (2009, 255) goes on to relate that "what had once been a city was [now] a weird lake" with "vile water" and "murky waters."

It left behind what Solnit (2009, 279) calls "a filthy, stinking, soggy mess." New Orleans was built in a swamp and it reverted to a swamp, except that it had been built in a native swamp and now it was a feral swamp (for the history of New Orleans as a wetland city, see Giblett 2016, chapter 11).

Just as Solnit's (2014b, 3) thinking about place has been refined and developed, so has it about disaster to become "broader and broader" to "now see much of our everyday life [as a disaster]—for its alienation and its destruction of souls and memory, as well as natural and social places." The "somewhatness" of the natural disaster of the storm was intensified and exacerbated by the unmitigated unnatural and cultural disaster of climate change and global warming (Solnit and Lutunatabua 2023). The unnatural disasters of nuclear bomb testing and climate change are caused by violence against the earth and the land. "Climate change is violence," as Solnit (2014b, 57) puts it bluntly in an essay of this title.

Like New Orleans, Paris was built in a wetland, a marsh in fact.[5] It is also a city like a female body too. Solnit (2000, 209) argues that Walter Benjamin undertakes the "transformation of city into female body." For Benjamin and the surrealists "Paris is a labyrinth [...] whose center is a brothel." The body is like a landscape and a construction, and vice versa. For Benjamin and the surrealists the female body is labyrinthine, and the female mind is too, an appropriate and affirmative metaphor as Solnit (2014b, 207) points out later as "labrys" has become "a lesbian icon." Not only is "the mind [...] also a landscape of sorts," or "can be imagined as a landscape," as Solnit (2000, 6; 2005a, 53) puts it, but also the landscape is a projection or extension of the mind. "The mind has mountains, cliffs of fall," as Gerard Manley Hopkins put it; cliffs of fall in mountainous landscapes are also of the mind.

Solnit (2000, 197, 198; 2005b, 99; 2007, 69, 70) calls Benjamin "one of the great scholars of cities and the art of walking them [...] who has become the patron saint of cultural studies." She also calls him "the incomparable, uncategorizable, German-Jewish essayist and theorist" and "the incomparable writer-philosopher" who "in his work was a historian, a theorist, a lyrical writer, creating writing as uncategorizable as it is influential."

By contrast, Solnit (2000, 183) simply and slightingly calls Raymond Williams "the literary historian." Calling him this is less than half the story, as Solnit points out with Benjamin who was equally not a literary historian

[5] For the history of Paris as a wetland city, see Giblett (2016, chapter 3).

too, though he dabbled in literary history along with philosophy and theology and sundry other things. Williams followed in the footsteps of Benjamin. Williams' *Culture and Society 1780–1950* published in 1958 is arguably the foundational text of Cultural Studies. His *The Country and the City* published in 1973 is arguably the foundational text of ecocultural studies and ecocriticism in general and political ecocriticism in particular. His later books, such as *Towards 2000* published in 1983, and his last essays collected in *Resources of Hope* published posthumously built on these foundations. Williams is "the patron saint of ecocultural studies" (Giblett 2020, chapter 10).

Williams is the incomparable, uncategorizable, Welsh essayist and theorist who, like Solnit, was an anti-nuclear activist. He regularly went on the annual "Campaign for Nuclear Disarmament" protest marches in London. He was also, like Solnit, a social justice activist. For instance, he helped deliver soup to striking Welsh miners in the 1970s, an unusual activity for a Cambridge University Professor. Williams was also one of the great scholars of the city and the country, communication and culture, technology and society, landscape and nature. He was also, like Solnit, an etymologist of these keywords in a book of this title and in some of his other books. He was also a theorist and activist for "the politics of hope," as Solnit (2005b, 148) calls it, with his "resources for a journey of hope" and for what he called "livelihood" in socialist ecology (see Giblett 2020, chapter 10).

13.5 Hoping, Walking, and Telling Stories

The mind not only has mountains, cliffs of fall, and vice versa, but the mind also has dismal swamps, sloughs of despond, and vice versa. Swamps, feral or native, can be considered not only as quaking zones where land and water meet but also as water-bodies where body and water meet. They are also places of the mind, especially of the uncanny, of horror and fascination. For Solnit (2005a, 180) "there is something fearful and mysterious about every body of water, murky water that promises unseen things in unseen depths, clear water that shows you the bottom far below as if you could fall into it." Both types of water, or both types of bodies of water, can be found in "nature" and "culture" whether it is the murky waters of native or feral swamps in which a monster (real or imagined) might lurk in their depths, or the clear water of an ocean rock pool or a chlorinated swimming pool.

Principally the native quaking zone is the swamp, or other wetland. It is also the soft and pliable centre of the human body. It applies to both the human body and the body of the earth, and even to the body of a book, such as when Solnit (1999, 111) refers to "the swampy midsection of [Thoreau's essay] 'Walking'," a slim volume, like Thoreau himself by all accounts. Swampy is an apt metaphor for the midsection of the body. The quaking zone of the swamp and the body is connected, not just by tropes, but by and in hope. Hope for Thoreau lives and thrives in the dark and dismal swamp and not in the manicured urban lawn. For Virginia Woolf (cited by Solnit 2005b, 1) "the future is dark" "with a darkness as much of the womb as the grave," Solnit (2005b, 6) adds. The swamp is a perfect place for generating what Solnit (2005b, 6) calls "hope in the dark" because its darkness is as much of the womb as the tomb. "Hope in the dark" for Solnit (2007, 169) is "hope in the future." It is "not too late" (Solnit and Lutunatabua 2023).

Walking connects places; walking is the activity that connects places. Driving, or using other forms of vehicular technology, transports the body from one place to another, whereas walking connects the body to places, with other places, and makes the connection between places. Walking for Solnit (2000, 9) nurtures "the sense of place that can only be gained on foot." Walking also connects the walker to their own body. Solnit reminds us that "walking in its rhythm and naturalness is the closest of all the acts we choose to the acts we don't [choose to do]: to breathing and the beating of our hearts, the other rhythms that direct the rhythm of walking" (Solnit 1999, 61). In less negative terms, Solnit (2000, 5) has also written "walking itself is the intentional act closest to the unwilled rhythms of the body, to breathing and the beating of the heart." Walking for Solnit (1999, 61) is "the only way to measure the rhythm of the body against the rhythm of the land." It is also a way to bring the rhythms of the body and the land together to move in time and place.

Walking is also a way to bring the body, land, and mind together. Walking for Solnit (2000, 5), "ideally, is a state in which the mind, the body, and the world are aligned." Walking connects the walker to their own mind when "the rhythm of walking generates a kind of rhythm of thinking" (Solnit 2000, 5). "Walking is natural" as Solnit (2000, 85) says, but the ways in which humans walk is cultural. There are fashions and styles in walking, as Marcel Mauss famously observed in his paper on "Body Techniques," just as much as there are fashions in clothes and hair. Walking is culturally political when it is marching, either to or in preparation for war, or for peace and against war of all sorts, such as globalization defined by Solnit (2005b, 66) as "war by other means." For Solnit (2005b, 81), "a march is when bodies speak by walking."

Bodies also speak by talking and telling stories. Solnit's progressive politics expressed in her writing 'entales' the telling of stories, not only of bodies and places, like cities such as New Orleans, and of landscapes, like Yosemite and Nevada, but also of everyday objects, such as her own desk. In her recent memoir Solnit tells the intertwined story of her desk and the friend who gave it to her, describing them in detail. A photograph of her desk is the frontispiece to her memoir (Solnit 2020b). Her desk bears the traces of two stories, hers and the givers, that are intertwined inextricably in the desk. Her desk holds memories, painful memories for her about her friend and her friend's own painful memories:

> [A] friend gave me a desk not long after I moved into the apartment, a woman's small writing desk or vanity, the one I am writing on now. It's a dainty Victorian piece of furniture, with four narrow drawers, two on each side and a broader central drawer above the bay in which the sitter's legs go, and various kinds of ornamentation—doweled legs, each with a knob like a knee, knobby ornaments, scallops on the bottom of the drawers, drawer pulls like tassels or teardrops. There are two pairs of legs on the front, two on the back, set beneath the side drawers. Despite all the frills, the old desk is fundamentally sturdy, an eight-legged beast of burden whose back has carried many things over the decades, or two beasts of burden side by side, yoked together by the desktop. The desk has moved with me three times. It's the surface on which I've written millions of words: more than twenty books, reviews, essays, love letters, several thousand emails to my friend Tina during the years of our near daily epistolary exchange, a few hundred thousand other emails, some eulogies and obituaries, including those of both of my parents, a desk at which I did the homework of a student and then a teacher, a portal onto the world and my platform for reaching out and for diving inward. A year or so before she gave me the desk, my friend was stabbed fifteen times by an ex-boyfriend to punish her for leaving him. She almost bled to death; she had emergency transfusions; she was left with long scars all over her body [...] she survived. [... Years later] she told me the full story, a story that could make your heart catch fire and the world freeze over. (Solnit 2020b, 43–44)

This desk is a body, or two bodies embracing and intertwining. A beast with two backs is a reference to a couple copulating in Shakespeare's

Othello (Act 1, Scene 1, ll. 126–127). From the two beasts of Solnit's desk, her writings were procreated and born, including the story of the desk, of the body of the previous owner, and of the brutal man who stabbed her almost to death, a truly terrifying story in its moral abhorrence and horrifying violence in its violation of the surfaces and depths of her body. This desk is not dead matter, but a living being with bodies attached and stories to tell. It is a talisman that tells of men's violence to women. The writer's desk can be a chamber of horrors whose exterior hides secrets and tells a story for the writer as Solnit's does. Telling stories is powerful. Storytelling, as Benjamin put it, is the messianic weak power through which the revolutionary now irrupts—as the storytelling of Rebeca Solnit shows and does with this and many other vital facets of progressive political ecology.[6]

References

Giblett, Rod. 2025. *Waking to Nature with Thoreau and Benjamin: Or, Psychopolitical Ecology*. New York: Bloomsbury.

Giblett, Rod. 1996. *Postmodern Wetlands: Culture History Ecology*. Edinburgh: Edinburgh University Press.

Giblett, Rod. 2008. *The Body of Nature and Culture*. Basingstoke: Palgrave Macmillan.

Giblett, Rod. 2009. *Landscapes of Culture and Nature*. Basingstoke: Palgrave Macmillan.

Giblett, Rod. 2011. *People and Places of Nature and Culture*. Bristol: Intellect Books.

Giblett, Rod. 2016. *Cities and Wetlands: The Return of the Repressed in Nature and Culture*. London: Bloomsbury.

Giblett, Rod. 2020. *New Lives of the Saints: Twelve Environmental Apostles*. Lanham, MD: Hamilton Books.

Giblett, Rod and Juha Tolonen. 2012. *Photography and Landscape*. Bristol: Intellect Books.

Giblett, Rod. 2026a. *Living with Nature in the Anthropobscene: Philosophy, Ecoculture, Political Ecology*. New York: Palgrave Macmillan.

Giblett, Rod. 2026b. *Wise Women for Loving Nature: Or, Femopolitical Ecology*. New York: Bloomsbury.

Solnit, Rebecca. 1999. *Savage Dreams: A Journey into the Landscape Wars of the American West*. Berkeley: University of California Press.

Solnit, Rebecca. 2000. *Wanderlust: A History of Walking*. London: Penguin.

[6] For further discussion of progressive political ecology, see Giblett (2025, 2026a, b).

Solnit, Rebecca. 2003. *As Eve Said to the Serpent: On Landscape, Gender and Art.* Athens, Georgia: University of Georgia Press.
Solnit, Rebecca. 2004. *River of Shadows: Eadweard Muybridge and the Technological Wild West.* London: Penguin.
Solnit, Rebecca. 2005a. *A Field Guide to Getting Lost.* New York: Viking.
Solnit, Rebecca. 2005b. *Hope in the Dark: The Untold History of People Power.* Edinburgh: Canongate.
Solnit, Rebecca. 2005c. The Uses of Disaster: Notes on Bad Weather and Good Government. *Harper's Magazine.* http://Harpers.Org/Archive/2005/10/0080774. Accessed 11 Sept 2008.
Solnit, Rebecca. 2007. *Storming the Gates of Paradise: Landscapes for Politics.* Berkeley: University of California Press.
Solnit, Rebecca. 2009. *A Paradise Built in Hell: The Extraordinary Communities that Arise in Disaster.* New York: Viking.
Solnit, Rebecca. 2010. *Infinite City: A San Francisco Atlas.* Berkeley: University of California Press.
Solnit, Rebecca. 2013. *The Faraway Nearby.* New York: Viking.
Solnit, Rebecca. 2014a. *Men Explain Things to Me.* Chicago: Haymarket Books.
Solnit, Rebecca. 2014b. *The Encyclopedia of Trouble and Spaciousness.* San Antonio, TX: Trinity University Press.
Solnit, Rebecca. 2017. *The Mother of All Questions.* Chicago: Haymarket Books.
Solnit, Rebecca. 2018. *Call Them by Their True Names: American Crises (and Essays).* Chicago: Haymarket Books.
Solnit, Rebecca. 2019a. *Cinderella Liberator.* Chicago: Haymarket Books.
Solnit, Rebecca. 2019b. *Whose Story Is This? Old Conflicts, New Chapters.* Chicago: Haymarket Books.
Solnit, Rebecca. 2020a. Climate Change, Covid—Our Hearts Ache. But a New Era Is Possible. We Can Do It. *The Guardian,* September 19. https://www.theguardian.com/commentisfree/2020/sep/19/covid-climate-change-. Accessed 2 Oct 2020.
Solnit, Rebecca. 2020b. *Recollections of my Nonexistence.* New York: Viking.
Solnit, Rebecca. 2021. The Violence at the Capitol Was an Attempted Coup. Call It That. *The Guardian*, January 7. https://www.theguardian.com/commentisfree/2021/jan/06/trump-mob-storm-capitol-washington-coup-attempt. Accessed 10 Jan 2021.
Solnit, Rebecca. 2025. *No Straight Road Takes You There: Essays for Uneven Terrain.* Chicago: Haymarket Books.
Solnit, Rebecca, and Joshua Jelly-Schapiro. 2016. *Non-Stop Metropolis; A New York City Atlas.* Berkeley: University of California Press.
Solnit, Rebecca, and Thelma Young Lutunatabua, eds. 2023. *Not Too Late: Changing the Climate Story from Despair to Possibility.* Chicago: Haymarket Books.
Solnit, Rebecca, and Rebecca Snedeker. 2013. *Unfathomable City: A New Orleans Atlas.* Berkeley: University of California Press.

CHAPTER 14

Conclusion: Nature and Environmental Writing in the Present and for the Future

Nature and environmental writing enable the conservation of places now and into the future in order to sustain their energy-producing processes in and as home habitats for humans and more-than-humans. They acknowledge and respect the role of these processes in the lives of more-than-humans. They enable the conservation of plants and animals, especially threatened and endangered species and their habitats, such as the Swallowtail Butterfly (depicted on the cover of this book) and its wetland home. They are prosaic in poetic prose, or prose poetry; they mix fact, fiction, and genres from heterogeneous sources; they are dialogic and polyphonic with nature and 'the environment' in the country and the city and between them; they are transdisciplinary with the sciences, arts, and humanities; they are minor literature written in major languages that deconstructs and decolonizes them and their cultures; they are carnivalesque about the grotesque and the monstrous and subvert hierarchies; and they engage femo- and psycho-political ecology and create political ecocriticism, including of nature and environmental writing.

Recent nature and environmental writing aim to halt and end the carboniferous hypercapitalist destruction by the fossil fuel industry of human and more-than-human beings and their homes. They signal the need for the total transition to renewable sources of energy that can help conserve these beings and recreate their homes. They develop new and practical ways of conserving and appreciating all biodiverse places for humans and

R. Giblett, *The Politics of Nature and Environmental Writing*, Transdisciplinary Environmental Humanities 1,
https://doi.org/10.1007/978-3-032-04427-3_14

more-than-humans. They continue the deconstruction and decolonization of the mainstream western cultural denigration of wetlands as places of horror and the monstrous, and its practices of dyking, draining, impounding, backfilling, embanking, and infilling them to create farmland and suburbia.

Recent nature and environmental writing also reconstruct environmental cultures so that all people (now and into the future) value, appreciate, and conserve places for humans and more-than-humans for what they provide for both and what they are themselves—places of exquisite and uncanny aesthetic qualities and vital living processes. They celebrate places in all their facets and bears witness to, speak truth to power about, their importance and value to and for life on earth and tell stories for doing so. They portray the exquisite nature of places so all people on earth appreciate and care for them more. They redevelop traditional ways and develop contemporary ways of being and living bio- and psycho-symbiotic livelihoods in all of the bioregional home-habitats of the living earth. Future nature and environmental writing will continue to do so too.

Index[1]

[1] Note: Page numbers followed by 'n' refer to notes.

R. Giblett, *The Politics of Nature and Environmental Writing*, Transdisciplinary Environmental Humanities 1,
https://doi.org/10.1007/978-3-032-04427-3

FSC
www.fsc.org
MIX
Papier | Fördert
gute Waldnutzung
FSC® C083411

Kolibri 360 GmbH
Ferdinand-Jühlke-Straße 7
99095 Erfurt, Deutschland
produktsicherheit@kolibri360.de